"十四五"职业教育部委级规划教材

 普通高等教育"十一五"国家级规划教材（高职高专）

PRACTICAL GUIDE TO GARMENT ENGLISH （FOURTH EDITION）

服装英语实用教材

（第4版）

张宏仁　梁娟　张小良　**主编**

陈孟超　蒋伟平　**副主编**

U0161779

课程宣传片

（扫描二维码观看视频）

中国纺织出版社有限公司

内 容 提 要

本书以一家从事服装加工的贸易公司为背景，内容基于服装外贸人才职业需求，划分为服装基础知识、外贸单据翻译、生产包装、外贸交流四个模块，覆盖该课程所有知识点和岗位技能知识点。按照教学内容的逻辑性、系统性、关联性、实战性和行业前瞻性原则，将资源碎片化和颗粒化，重构资源体系，配套46个精品视频课及教学PPT。本书可作为应用英语（服装外贸方向）、服装设计与工艺、服饰设计的专业教材。

本书选材新颖、内容丰富、实用性强。让读者在学习服装专业英语的同时，了解服装专业知识，掌握服装外贸流程，即学即用，增强职业竞争力。

图书在版编目（CIP）数据

服装英语实用教材 / 张宏仁，梁娟，张小良主编；陈孟超，蒋伟平副主编. -- 4 版. -- 北京：中国纺织出版社有限公司，2023.11

"十四五"职业教育部委级规划教材

ISBN 978-7-5229-0885-4

Ⅰ. ①服… Ⅱ. ①张… ②梁… ③张… ④陈… ⑤蒋… Ⅲ. ①服装 - 英语 - 职业教育 - 教材 Ⅳ. ① TS941

中国国家版本馆 CIP 数据核字（2023）第 161867 号

责任编辑：宗 静 亢莹莹 特约编辑：周 蓓
责任校对：高 涵 责任印制：王艳丽

中国纺织出版社有限公司出版发行
地址：北京市朝阳区百子湾东里A407号楼 邮政编码：100124
销售电话：010—67004422 传真：010—87155801
http://www.c-textilep.com
中国纺织出版社天猫旗舰店
官方微博 http://weibo.com/2119887771
三河市宏盛印务有限公司印刷 各地新华书店经销
2000年8月第1版 2007年11月第2版 2015年1月第3版
2023年11月第4版第1次印刷
开本：787×1092 1/16 印张：15
字数：320千字 定价：59.80元

第4版
前言

在产业集群发展趋向品牌化、规模化、国际化的产业转型升级大背景下，既熟悉服装生产与经营，又熟练掌握服装专业英语的人才，成了服装企业急需的"香饽饽"。

为了适应目前的经济情况，为服装行业培养更多的综合型人才，笔者根据自己在合资企业多年的实践经验和服装专业英语教学的丰富经验，精心编写了本书。为了方便读者在有限的时间内，更有效地学习到实用的服装专业英语，本书以一家从事服装加工的贸易公司为背景，以实际生产运作为主线，用英文的形式（配参考译文）介绍该服装公司的部门划分、岗位职责和相关外贸工作过程，其中包括样品制作与确认、加工/采购合同的签订、外贸单据的处理、面辅料、质量控制、服装包装和标志、服装外贸函电、服装电子商务等，可即学即用，增强职业竞争力。在内容上，本书展示了涉外服装企业的生产运作模式，介绍了服装专业英语知识（例如，选取了许多服装企业的实际单据和日常公务表格），让读者在学习专业英语的同时，也掌握服装加工贸易企业的运作流程。

本书第2版为普通高等教育"十一五"国家级规划教材(高职高专)，获中国纺织服装教育学会"纺织服装教育'十一五'部委级优秀教材"。该教材自2000年出版以来，受到了服装专业师生和服装业界人士的一致好评，2007年再版，2015年第3次改版，在广大读者中影响深远。随着我国教育教学改革的不断深入和科学技术的飞速发展，部分教材内容已经稍显陈旧，需要更新。基于此，本书更新了服装生产和外贸的实用文体、服装电子商务、时装快讯等，增加了色彩在服装中的应用、服装部位名称、服装样衣、服装英语的缩写、服装设备和服装包装和标识等章节，编写出该书的第4版。该书既可作为高等职业院校的专业英语教材，也可供其他职业培训学校以及服装从业人员学习、参考。

教材由张宏仁、梁娟、张小良任主编，陈孟超、蒋伟平任副主编，刘书晴、卢凤仪、邓沙琪、胡颖珊、龚凤卿、王琰、李艳辉参与编写。由于编写时间较紧，编者水平有限，不足之处在所难免，还请有关专家、学者予以指正。

编者
2023年4月

第3版
前言

在产业集群发展趋向品牌化、规模化、国际化的产业转型升级大背景下，既熟悉服装生产与经营，又熟练掌握服装专业英语的人才，成了服装企业急需的"香饽饽"。

为了适应目前的经济情况，为服装行业培养更多的综合性人才，笔者根据自己在合资企业多年的实践经验和服装专业英语教学的丰富经验，精心编写了本教材。为方便读者在有限的时间内，更为有效地学习到实用的服装专业英语，教材以一家从事服装加工的贸易公司为背景，以实际生产运作为主线，用英文的形式（配参考译文）介绍了该服装公司的部门划分、岗位职责和生产运作过程，其中包括样品制作与确认、加工／采购合同的签订、纸样制作、裁剪、缝制、生产线平衡、质量控制、相关的工艺、市场销售技能、技术实用资料，可即学即用，增强职业竞争力。在内容上，本教材展示了涉外服装企业的生产运作模式，介绍了服装企业常用的英语知识（例如选取了许多企业的实际单据和日常公务表格），让读者在学习专业英语的同时，也掌握了服装加工贸易企业的生产运作流程。

本书的上一版《服装英语实用教材》（第二版）为普通高等教育"十一五"国家级规划教材（高职高专），获中国纺织服装教育学会"纺织服装教育'十一五'部委级优秀教材"。该教材自2000年出版以来，受到了服装专业师生和服装业界人士的一致好评，2007年再版，在广大读者中影响深远。随着我国教育教学改革的不断深入和科学技术的飞速发展，部分教材内容已经稍显陈旧，需要更新。基于此，本书更新了服装生产和外贸的实用文体、时装快讯等，增加了服装电子商务章节，编写出该书的第三版。该书可作为高等职业院校的专业英语教材，也可供其他职业培训学校以及服装从业人员学习、参考。

教材由张宏仁任主编，梁娟、张小良任副主编。由于编写时间较紧，编者水平有限，不足之处在所难免，还请有关专家、学者予以指正。

编者
2014年4月

第2版
前言

中国加入WTO以后，越来越多的中国服装企业参与到国际竞争中。在这种情况下，既熟悉服装生产与经营，又能熟练掌握服装专业英语的人才，成了服装企业急需的"香饽饽"。

为了适应目前的经济情况，为服装行业培养更多的综合人才，笔者根据自己在合资企业多年的实践经验和服装专业英语教学经验，精心编写了本教材。为方便读者在有限的时间内，更为有效地学习到实用的服装专业英语，本教材以一家从事服装加工的贸易公司为背景，以实际生产运作为主线，用英文的形式（配参考译文）介绍了该服装公司的部门划分、岗位职责和生产运作过程，其中包括样品试制与确认、加工/采购合同的签订、纸样制作、裁剪、缝制、生产线平衡、质量控制、包装、服装市场营销、沟通英语、服装跟单和商务信函等方面的知识和相关的工艺、技术实用资料，能让读者即学即用，增强职业竞争力。在内容上，本教材展示了涉外服装企业的生产运作模式，介绍了服装企业常用的英语知识（如选取了许多企业的实际单据和日常公务表格），让读者在学习专业英语的同时，也掌握了服装加工贸易企业的生产运作流程。

本教材由张宏仁主编，张小良副主编，参加编写的人员还有谭雄辉、梁娟等，全书由张宏仁统稿与审定。

本书是普通高等教育"十一五"国家级规划教材（高职高专）之一，为大专院校、职业大学服装专业的适用教材，亦可供服装从业人员学习、参考。

由于编写时间较紧，加之编者水平有限，不足之处在所难免，尚祈有关专家、学者给予指正。

编者
2007年7月

第1版
前言

目前，我国服装加工行业制单中，有相当部分是外来加工单（如来自美国、加拿大及欧洲一些国家等）。在制单中常有服装专业英语词汇，或整份皆以英文填制。中专服装专业毕业生将是制衣业生产第一线技术骨干。如目前毕业的中专生很多在制衣企业中充当QC、打纸样、制订工艺流程、小组长、车间主任、技术主管、营销等角色，都要接触到制单。为适应形势的需要，中专服装专业学生很应该具备一定服装专业英语的阅读、书写和会话能力。为方便学生在有限的时间内，更有效地学习到将来工作中非常实用的专业英语，主编者根据自己在合资企业多年的实践工作经验和多年的服装专业英语教学经验，精心整编本教材。本书在一些名词术语英译汉时，既有书面译法，又有广东话译法，使本书既方便内地学生学习，又方便沿海地区（接触"港式制单"）的学生学习。

编者多年的教学实践证明，本书可作为职业学校服装专业的英语教材。本书主要以制衣业生产流程为主线，紧密联系工厂实际运作，选取了一些实际制单为题材的课文。这是在校学生最为缺乏的知识点。通过本书的学习，让学生既学到专业英语，又可了解到制衣企业各岗位之职责，学到一些工种的实际操作方法，增长专业知识，增强学生毕业求职的竞争力。

本书专业词汇较多，最好安排在基础英语课程结束之后进行教学。按教学计划这门课的总学时为80学时，其中的学时分配：讲课72学时，考试4学时，机动4学时。

本书由广东省纺织工业学校张宏仁主编，参与本书编写的有张宏仁（负责编写第1至第8课课文及译文、第9课一二部分课文及译文，第10至第12课课文及译文，第14至第16课课文及译文以及附录Ⅰ、附录Ⅲ、附录Ⅴ、附录Ⅵ、附录Ⅷ和附录Ⅹ）、张小良（负责编写第3课第二部分及译文，第9课第二部分的部分表格及译文，第13课课文及译文，第18课课文及译文，附录Ⅲ和附录Ⅳ、谭雄辉（负责编写第17课课文及译文和附录Ⅴ），全书由张宏仁统稿和审定。

本书可供职业学校服装专业作为专业英语教材用书，亦可供服装行业从业人士学习、参考以及经贸外语专业的师生阅读。

由于编者水平有限，不妥之处，尚祈各地师生、各位读者不吝指教，不胜感激。

编者
2000年3月

教学内容及课时安排

课程性质（课时）	章（课时）	节	课程内容
模块一 服装基础知识 （16课时）	第一章 （2课时）		• 我们的公司
		一	公司组织架构
		二	岗位职责划分
		三	服装外贸跟单
	第二章 （6课时）		• 面料
		一	面料纤维
		二	面料分类
		三	常用服装面料
	第三章 （6课时）		• 辅料
		一	里料
		二	衬料
		三	垫料
		四	填料
		五	缝纫线
		六	扣紧材料
		七	其他辅料
	第四章 （2课时）		• 色彩在服装中的应用
		一	色彩基本知识
		二	常用色彩的英文表达
模块二 服装外贸单据翻译 （22课时）	第五章 （6课时）		• 服装部位名称
		一	西装外套各部位的中文表述
		二	夹克各部位的中文表述
		三	裤子各部位的中文表述
		四	连衣裙各部位的中文表述
	第六章 （4课时）		• 服装样衣
		一	样衣的分类
		二	服装的尺码及标注

续表

课程性质（课时）	章（课时）	节	课程内容
模块二 服装外贸单据翻译 （22课时）	第七章 （6课时）		• 样板单和确认样板卡
		一	样板单
		二	确认样板卡
	第八章 （6课时）		• 加工合同和采购订单
		一	加工合同
		二	采购订单
模块三 服装生产和包装 （22课时）	第九章 （4课时）		• 服装英语的缩写
		一	缩写的目的
		二	服装外贸缩写
	第十章 （2课时）		• 服装设备
		一	测量工具
		二	描图工具
		三	熨烫工具
		四	缝纫工具
		五	裁剪工具
	第十一章 （6课时）		• 生产通知单
		一	上装的生产通知单
		二	下装的生产通知单
	第十二章 （6课时）		• 质量控制与检验报告
		一	检验方法
		二	质量标准
		三	测量指南
		四	服装疵点的表述
		五	检验报告
	第十三章 （4课时）		• 服装包装和标识
		一	服装包装功能
		二	服装包装分类
		三	服装标识
		四	运输标志
		五	装箱单

课程性质（课时）	章（课时）	节	课程内容
模块四 服装外贸交流 （12课时）	第十四章 （3课时）		• 商业电子邮件
		一	商业电子邮件简介
		二	商业电子邮件格式
		三	服装外贸商业电邮案例
	第十五章 （3课时）		• 服装外贸中的沟通交流
		一	服装产品的沟通交流
		二	样衣的沟通交流
		三	谈判和签订合同的沟通交流
		四	质量和投诉的沟通交流
	第十六章 （3课时）		• 服装电子商务
		一	服装电子商务
		二	服装电子商务平台
		三	如何提升外贸业绩
	第十七章 （3课时）		• 时装快讯
		一	香云纱
		二	元宇宙与虚拟时尚
		三	GUCCI春/夏系列时装秀
		四	内衣元素

注 各院校可根据自身的教学特点和教学计划对课程时数进行调整。

配套微课资源索引

续表

目录

Module 1 Basic Garment Knowledge
模块一 服装基础知识

Module 2　Garment Foreign Trade Documentary Translation
模块二　服装外贸单据翻译

Module 3　Garment Producing and Packaging
模块三　服装生产和包装

Module 4　Communication in Garment Foreign Trade
模块四　服装外贸交流

Module 1
Basic Garment Knowledge

模块一
服装基础知识

Basic Garment Knowledge
服装基础知识

Chapter 1　　Our Company
第一章　　我们的公司

项目名称：我们的公司

项目内容：1．公司组织架构

　　　　　2．岗位职责划分

　　　　　3．服装外贸跟单

教学学时：2课时

教学目的：让学生了解服装外贸企业的业务范围、组织架构、岗位设置和职责划分、服装外贸跟单员的业务内容和流程等。

教学方式：由教师通过课前导入任务引出本章内容。结合案例分析讲解公司架构、岗位职责和服装外贸跟单员的具体职责。观看拓展视频，并总结归纳学习难点和重点，学生进行实操练习。

教学要求：（1）掌握本课词汇。

　　　　　（2）熟悉服装企业各岗位基本职责及相互间关系。

　　　　　（3）熟练掌握服装外贸跟单员的业务内容和流程。

Chapter 1   Our Company

Our company is a garment manufacturing and trading limited company. Our customers are mainly from U.S.A, Canada, Northern Europe and Eastern Europe. The business scope of our company is all kinds of woven garments and other fabric garments for men' s, women's and children's wear such as shirts, pants, pajamas, jacket, coat, sports suits and skirt.

There are mainly seven departments who carry out the routine work to ensure the smooth running of our company. The following Fig.1-1 shows the duty divisions in the organization of our company, and Table 1-1 shows job titles and job descriptions in this company.

Part 1   Organization Chart

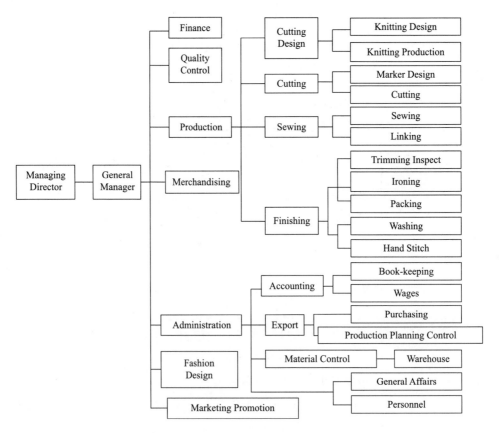

Fig.1-1   Organzation Chart

Part 2 Job Titles and Job Descriptions

Table 1-1 Job Titles and Job Descriptions

Job Titles	Job Descriptions
Technologist Level	
Production Manager	Plans, organizes, directs and controls all aspects of production, including the co-ordination of associated areas to ensure the most effective and economical means of production
Quality Control Manager	Plans, organizes, directs and controls quality control procedures in all stages of production to ensure incoming materials and products to comply with required standards and specifications
Fashion Designer	Creates designs for garment products harmonizing aesthetic considerations with technical, merchandising and costing requirements
Merchandising Manager	Keeps abreast of the up-to-date fashion and quality requirements of the markets; plans, approves and supervises the co-ordination and presentation of sampling and quotations, negotiates with buyers overseas and follows up of buyers' order. Allocates the tasks, checks and implements the quality control and inspection of each order reasonably. Liaises with appropriate departments to ensure the prompt shipment of customers' orders
Technician Level	
Production Department Supervisor	In charge of the production department, plans and controls production efficiently and assists in production schedule and work study
Production Section Supervisor	In charge of a section in the production department, including directing preparation work prior to production, controlling production quality and allocating work to each machine
Quality Inspection Supervisor	In charge of a section in the quality inspection department, inspects the quality of products in all stages of processing and maintains the quality standard of finished products
Pattern Maker	Designs and makes patterns for various parts of a garment and whole garments
Knitwear Technician	Designs and writes knitting instructions according to drawings, specifications, designs or ideas for hand and power operated knitting machine operators and for trimmings
Merchandiser	In charge of pre-sales and after-sales tracking of foreign trade orders, production orders, progress tracking, quality confirmation, transportation and logistics, etc.
Craftsman Level	
Garment Machine Mechanic	Installs, converts, overhauls, maintains, repairs, designs and makes attachments for sewing machinery
Knitting Machine Mechanic	Installs, converts, overhauls, maintains, repairs, designs and makes attachments for knitting machinery
Pattern Grader/ Marker maker	According to size specifications, produces full ranges of different sizes of patterns from master patterns, designs marker lays for production orders
Garment Operator Instructors	To train trainees for one or more jobs at operative level. Also to retrain and further train existing operatives in new and existing skills
Plant Maintenance Mechanic	Maintains, overhauls and repairs all electrical and mechanical equipments, including small motors, electrical mechanical hand tools and associated control equipments, and also maintains simple boiler and pressing equipment under the direction of a supervisory grade of employee

Continued

Job Titles	Job Descriptions
Operative Level	
Inspection Operative	Inspects materials, fabrics, garment parts and garments for faults and quality
Cutter (Garment)	Cuts cloth into parts of garments by hand or machine from the marker lay plan
Cloth Spreader	Spreads cloth into layers to facilitate cutting
Lockstitch S/M Operator (Garment)	Operates a lockstitch S/M for sewing the component parts of garments
Special Purpose S/M Operator (Garment)	Operates a special purpose sewing machine for sewing the component parts of garments
Make-through Operator (Garment)	Operates S/M for assembling garments on the make-through systems
Hand Knitting M/C Operator (knitwear)	Operates hand knitting machines to produce garment parts
Power Knitting M/C Operator (knitwear)	Operates power knitting machines to produce garment parts
Linking Machine Operator (knitwear)	Links Knitted garments parts by means of linking machine
Hand Stitcher (knitwear)	Stitches by hand garment parts and trimmings
Mender (knitwear)	Mends by hand defective parts in knitted fabric
Presser	Presses sewn garments and parts by hand iron or pressing machine
Unskilled Level	
General Worker	Unskilled workers who may undertake the work of a trimmer, a cone winder, a packer or a cleaner, etc.

Part 3   Garment Foreign Trade Documentary

As a garment foreign trade enterprise, the merchandising department is one of the most important departments of the company. Garment foreign trade documentary has the characteristics of many business links, wide work range, strong professionalism, fast pace, changeability and so on. Since the trade contract is signed, garment merchandiser will track or operate some or all links such as garment processing, shipment, insurance, inspection, customs declaration and foreign exchange settlement according to the contract and relevant documents, and assist in the performance of the trade contract. As a follower of foreign trade orders, the professional ability and quality of garment foreign trade merchandisers affect the service quality and image of the company, and even the survival and development of the enterprise directly (Fig.1-2).

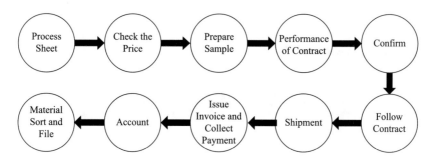

Fig.1-2　Work Content and Process of Garment Foreign Trade Documentary

New Words and Expressions

1. negotiate　谈判；磋商
2. liaise　（口）做联络人；联络
3. prior to　在……之前
4. install　装设（器械）
5. convert　改装；改造
6. overhaul　细密检查；检修
7. attachment　附件
8. trainee　受训者；（通常指接受工艺训练的）练习生
9. boiler　锅炉；汽锅
10. facilitate　使容易；使便利
11. lockstitch S/M (sewing machine)　平车；平缝机
12. make-through　单件制作（指一件服装全由一个人缝制）
13. mend　织补；缝补
14. trimmer　剪线工
15. cone　锥体；锥形；宝塔筒子
16. winder　（缝纫机）绕线器
17. packer　包装工人
18. associate　有联系的；有关的
19. supervisory　监督的；管理的
20. employee　雇员；受雇者
21. oversee　监督；视察
22. scope　活动或观察的范围
23. pajama　睡衣
24. carry out　实行完成
25. running　运转
26. division　分开；划分
27. knitting　编织；编织物
28. marker　排料图；唛架
29. linking　连接；接合；缝口
30. trimming　装饰；装饰物；修剪
31. administration　经营；管理；行政部门
32. book-keeping　簿记；记账
33. material　物料；面料
34. affair　事务；业务
35. promotion　以广告推销；促销
36. co-ordination　协调；配套
37. incoming　进来的；来到的
38. specification　规格
39. harmonizing　使调和；符合
40. aesthetic　美学原理；审美论
41. keep abreast of　与……并进；不落后
42. up-to-date　新近的；时髦的
43. merchandiser　跟单员

Exercises

List the detail job descriptions of merchandiser (foreign trader) in English.

导入：在当前形势下，如何管理服装外贸企业？

第一章　我们的公司

我们公司是一家服装加工和贸易有限公司。客户主要来自美国、加拿大、北欧和东欧。公司业务范围包括各类机织和其他面料的男装、女装和童装，诸如衬衣、裤子、睡衣、夹克、外套、运动衣和裙子。

公司主要有七个部门负责日常工作以确保公司顺畅运行。图1-1列出了我们公司组织架构的职责划分，表1-1列出了工作类别和工作简述。

第一节　公司组织架构

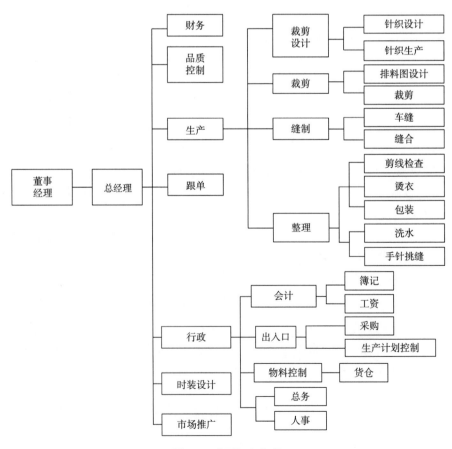

图1-1　公司组织架构

第二节　岗位职责划分

表1-1　工作类别和工作简述

工作类别	工作简述
技师级	
生产经理	策划、编排、指导及控制生产方面的各项工作，包括协调其他相关工作，以确保实施高效、经济的生产手段
质量控制经理	策划、编排、指导及控制生产各阶段的品质管理程序，以确保入厂的原料及产品符合既定标准的规格
时装设计师	设计时装，以和谐美观为原则，同时符合技术、销售及制作成本的要求
跟单经理	不断关注最新的时装趋势及市场对品质的需求，策划、批准及监督样品制作与报价之间的配合及提交等工作，并与客户商谈、统筹及处理客户订单，合理分配跟单员的工作任务，检查和落实每个订单的质量控制和检验工作，联络相关部门以确保客户订单能依期付运
技术员级	
生产部总管	管理生产部，有效地计划及控制生产，并协助编排生产程序表和工作研究
生产组管理员	管辖生产部门内一个小组，包括指导生产前的准备工作，管制生产质量及分派工作
品质检查管理员	负责管理品质检查部门内一个小组的工作，检查生产过程中各阶段的产品质量，并确保制成品符合质量标准
纸样设计员	设计及绘制各类部件及其整件服装的纸样
针织技术员	根据绘图和规格来设计或编写针织工作说明，为手动与电动针织机操作员及修剪工艺提供设计与构想
跟单员	负责外贸订单的售前售后跟踪、生产下单、进度跟踪、质量确认、运输物流等
技工级	
制衣机械工	担任制衣机械的安装、改装、大修、保养、修理、设计及制作附件等工作
针织机械工	担任针织机械的安装、改装、大修、保养、修理、设计及制作附件等工作
放样员/排料员（唛架员）	按照尺码规格，根据基础样板绘制各种不同尺码的纸样，依据订单，设计排料图以供生产
缝制指导工	训练半技工程度的练习生担任一项或多项工作以达到操作工的水平。并负责向现职操作工提供继续教育和再培训，使其获得新技能
设备保养技工	在管理级人员的指导下保养、大修与修理各种机电设备，包括小型电动机、电动机械手工具及相关的控制设备，并担任简单的锅炉及整烫设备的保养工作

续表

工作类别	工作简述
操作工级	
检查工	检查物料、布料、服装各部分及成衣的瑕疵点和质量
裁剪工（服装）	以手剪或使用机器依照排样单将布料裁剪成服装各部件
铺布工	将布料拉直铺叠成层，以便裁剪
平缝机车工（服装）	运用平缝机（平车）车缝服装各部件
特种缝纫机（服装）	操作特种缝纫机（衣车），车缝服装各部件
全件制车工（服装）	运用缝纫机（衣车）缝合全件服装
手动针织机织工（针织品）	操作手动针织机，织制服装各部件
电动针织机织工（针织品）	操作电动针织机，织制服装各部件
缝纫工（针织品）	运用套口机（缝盘机）缝连针织服装各部件
挑缝工（针织品）	以手工挑缝服装各部件及装饰品
补衣工（针织品）	以手工织补针织品的破损部份
整熨工	使用熨斗或熨衣机烫平服装及其附属部件
非技工级	
杂工	担任剪线头、打毛、包装或清洁等非技术性工作

第三节　服装外贸跟单

作为服装外贸企业，跟单部是公司最重要的部门之一。服装外贸跟单具有业务环节多、工作覆盖面广、专业性强、工作节奏快、变化多等特点。在贸易合同签订后，服装外贸跟单员依据合同和相关单证对服装加工、装运、保险、报检、报关、结汇等部分或全部环节进行跟踪或操作，并协助履行贸易合同。作为外贸订单的跟进者，服装外贸跟单员的专业化能力和素质，直接影响公司的服务品质和企业形象，甚至是企业的生存与发展（图1-2）。

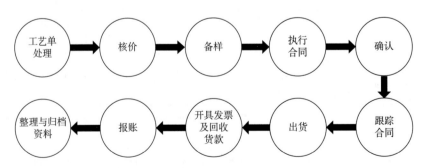

图1-2　服装外贸跟单工作内容及流程

Basic Garment Knowledge
服装基础知识

Chapter 2　Fabric
第二章　面料

项目名称：面料

项目内容：1. 面料纤维

2. 面料分类

3. 常用服装面料

教学学时：6课时

教学目的：让学生了解服装面料的基本概念、分类和特点，并培养学生服装面料专业术语英文表达能力。

教学方式：由教师通过课前导入任务引出本章内容。结合案例分析讲解纺织纤维分类，服装材料的分类及特点。观看拓展视频，了解机织和针织面料区别。结合外贸案例，深入学习常用服装面料的英文表述。总结归纳学习难点和重点，学生进行实操练习。

教学要求：（1）掌握本课词汇。

（2）了解纺织纤维分类及特点。

（3）熟悉服装面料的分类和特点。

（4）熟练掌握常用服装面料的英文表述。

Chapter 2　Fabric

Fabric refers to the material used to make garment. As one of the three elements of garment, fabric can not only interpret the style and characteristics of garment, but also directly affect the performance effect of garment color and shape.

Part 1　Textile Fibers

The textile fibers can be classified into two groups: natural and man-made fibers. Natural fibers can further be classified into vegetable, animal and mineral fibers.

"Textile" originally is applied to woven fabric, now the term is applied to any manufactured fibers, filaments, or natural and man-made yarns, obtained by interlacing (the process of forming a fabric by the interlacing of warp and weft), such as threads, cords, ropes, braids, laces, embroideries, nets, and garment made by weaving, knitting, felting, bonding, tufting, etc.

To understand the knowledge of the classification about textile fibers is the first step in the study of fabric. Detailed fibers may be classified according to their original characteristics and generic types (Fig.2-1).

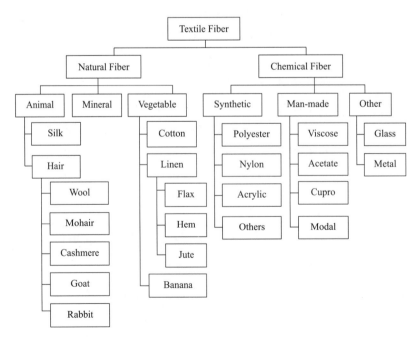

Fig.2-1　Classification of Textile Fibers

Natural fibers have different properties and are generally expensive, some of which are expensive. Cotton fiber is the most important natural fiber because it accounts for half of the fiber applications in the world. Hemp is another important natural cellulose fiber, but the production cost is much more expensive than cotton. The most important animal fibers are wool, mohair and cashmere, which can be used to make suits, coats, skirts, knitted garment, carpets, etc. Another kind of animal fiber is silk fiber, which is generally expensive and can be used to make women's blouses or shirts.

Compared with natural fiber, chemical fiber is cheaper than natural fiber and widely used, but its performance is poor. Polyester fiber is an important chemical fiber which is second only to cotton fiber. In the garment industry, nylon is produced earlier than polyester fiber. Acrylic fiber also shares an important market in the knitted garment industry because of its low price and machine washable. Man made fiber is generally produced by dissolving natural cellulose, chemical treatment and mechanical processing, so it is also called "regenerated" fiber. Man made fibers are widely used in the textile and garment industry as lining, coveralls, long skirts and even household products.

Part 2　Classification of Fabrics

For the classification of fabrics, the following two classification standards are usually adopted.

According to the standard of fabric structure, it is divided into woven fabric, knitted fabric and non-woven fabric. Woven fabrics and knitted fabrics are the two most common types of garment fabrics.Non-woven fabric is processed by bonding, fusion or other mechanical and chemical methods. For example: shopping bags.

Woven fabric is a fabric formed by interweaving warp and weft vertically. The warp yarn is called warp and the weft yarn is called weft. Woven fabric is generally divided into plain, twill and satin weave.The advantages of are stable structure, not easy to change, flat surface, easy to be treated with printing and jacquard treatment.The disadvantages of woven fabric is poor elasticity. Commonly used woven fabrics mainly include chiffon, Oxford, denim, twill, flannel, damask, etc.Woven fabrics are often used in coats, suits, jeans, trousers and so on(Fig.2-2).

Knitted fabric is a fabric formed by the bending of yarn in space.Each coil is composed of a yarn. The yarn or filament is formed into a coil with a knitting needle, and then the coils are sleeved in series with each other to form a coil structure. Knitted fabrics are divided into warp knitted fabric and weft knitted fabric. Knitted fabrics are divided into single-sided and double-sided.The advantages of knitted fabrics are good scalability, good softness, good moisture absorption and air permeability, good wrinkle resistance. The disadvantages of knitted fabrics are easy to separate, poor dimensional stability and easy snagging and fuzzing.Common used knitted fabrics mainly include single jersey, velour, bird's-eye, mesh fishnet, etc. Knitted fabrics are usually used in T-shirts, T-Shirts, sportswear, sweaters and so on (Fig.2-3).

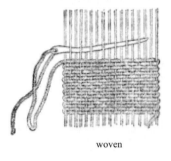
woven

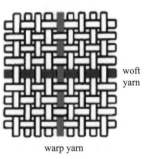
woft yarn
warp yarn

Fig.2-2　Structure of Woven Fabric

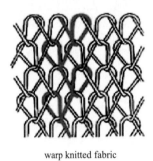
warp knitted fabric

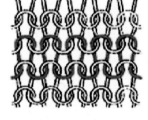
weft knitted fabric

Fig.2-3　Structure of Knitted Fabric

According to the standard of textile fiber composition, it is divided into natural fabric, chemical fabric and blending fiber.

The cost of natural fabrics is high, and the cost of chemical fabrics is low, but the performance is low. Therefore, most of the fabrics on the market are blended products. Blending can reduce the cost of products and improve the performance of products.

Part 3　Commonly Used Garment Fabrics

In the garment industry, the commonly used garment fabrics mainly include: cotton, linen，silk, wool, leather, polyester, spandex, polyamide, acrylic and modal etc.

1　Cotton

Cotton is a fabric produced by textile technology with cotton as raw material.The advantages of cotton are light and soft, warm, good moisture absorption, good permeability. The disadvantages of cotton are easy-shrinking, easy-wrinkling,ironing frequently. It is often used to make fashion, casual wear, underwear, and shirts.

2   Linen

Linen is a fabric made of fibers obtained from various linen plants.Generally, there are ramie,flax,jute,etc.The advantages of linen are good intensity,heat-conduction,moisture absorption.The disadvantages of linen are hard and poor hand-feeling. It is generally used to make casual garment and work garment. At present, it is also used to make ordinary summer garment.

3   Silk

Silk belongs to protein fiber and is the most textured natural fiber. The advantages of silk are light weight,soft and smooth,elegant and luxury,colorful,good moisture absorption. The disadvantages of silk is crease and fading easily. It can be used to make all kinds of high-grade garment, especially suitable for women's garment.

4   Wool

Wool belongs to protein fiber, which is the general name of all kinds of wool, cashmere, rabbit hair, camel hair and other fabrics. The advantages of wool are wearable,soft,elegant,elastic and warm.The advantages of wool is difficult to wash.It is suitable for making formal dress, suit, coat and other formal and high-grade garment.

5   Leather

Leather is a general term for pig leather, cow leather, sheep leather, horse leather, donkey leather, kangaroo leather, crocodile leather, ostrich leather, etc.The advantages of leather are warm and graceful. The advantages of leather are high-price and hard-storage. It is generally used to make fashion and winter wear.

6   Polyester

Polyester is an important variety of synthetic fiber, and the price is relatively cheap. Polyester is widely used in the manufacture of garment fabrics and industrial products.The advantage of polyester are not easy to shrink，not easy to fade，good heat resistance and good wrinkle resistance. The disadvantage of polyester are poor moisture absorption, poor permeability, easy to generate static electricity,poor hot melting/ fusibility.

7   Spandex

Spandex is translated as "Spandex", which is an elastic fiber, abbreviated as PU. It has the advantage of good elasticity and can be elongated by 6-7 times. The disadvantage of spandex is poor moisture absorption.

8   Polyamide

Polyamide namely nylon, is the first synthetic fiber in the world. Its appearance makes the appearance of textiles take on a new look. Its synthesis is a major breakthrough in the synthetic fiber industry.The advantage of polyamide are not easy to shrink,not easy to fade, good wrinkle resistance and good elasticity. The disadvantage of polyamide are poor permeability, poor moisture absorption, poor heat resistance and easy to generate static electricity.

9　Acrylic

Acrylic also known as polyacrylonitrile. The performance of acrylic fiber is very similar to wool, with good elasticity. When the elongation is 20%, the rebound rate can still be maintained at 65%. It is fluffy, curly and soft, and its warmth retention is 15% higher than that of wool. It is known as synthetic wool.The advantage of acrylic are good elasticity,good sun resistance, good warmth retention and acid and oxidant resistant. The disadvantage of Acrylic are poor alkali resistance, poor moisture absorption and poor hot fusibility.

10　Modal

Modal is a kind of cellulose fiber, which is pure man-made fiber. It is made of wood slurry made from shrubs produced in Europe and then made through special spinning process. Fibers can be blended and interwoven with a variety of fibers, such as cotton, hemp and silk, which are mostly used in the production of underwear.The advantage of Modal are soft, slippery, bright, good drape, moisture absorption and good color solid.The disadvantage of modal is poor rigidity.

With scientific and technological innovation, pineapple, banana, apple and coffee, which are the most common foods in daily life, have been widely used as sustainable natural textile raw materials in the fashion industry.

New Words and Expressions

1. textile fiber　纺织纤维
2. man-made fiber　人造纤维
3. regenerated fiber　再生纤维
4. natural fiber　天然纤维
5. mineral fiber　矿物质纤维
6. woven fabric　机织布料
7. filament　长纤丝
8. yarn　纱线
9. interlacing　交织
10. warp　经纱
11. weft　纬纱
12. cord　绳线
13. rope　绳索
14. braid　织带
15. lace　花边
16. embroidery　绣制品
17. net　网布
18. weaving　机织
19. knitting　针织
20. felting　制毡
21. bonding　毡合
22. tufting　起毛
23. synthetic fiber　合成纤维
24. polyester　聚酯
25. polyamide　聚酰胺；锦纶
26. linen　麻
27. flax　亚麻
28. hemp　大麻
29. silk　丝
30. wool　羊毛
31. mohair　马海毛
32. cashmere　羊绒

33. jute　黄麻
34. nylon　尼龙
35. rayon　再造纤维
36. acetate　醋酯纤维
37. blending　混纺
38. possess　有；具有
39. disadvantage　缺点
40. preshrink　预缩水
41. shrinkage　缩水率
42. characteristic　特性；特点
43. shiny　有光泽的；发亮的
44. luxurious　华丽的；舒适的
45. appearance　外表
46. slippery　光滑的；湿滑的
47. unravel　解开……的线；拆开
48. property　特性；属性；性质
49. stiffness　坚硬度
50. luster　光泽；光彩
51. crease　折印；折痕；皱纹；褶裥；裤缝
52. lustrous　有光泽的；光辉的
53. acetate　醋酸
54. performance　性能
55. composition　成分
56. moisture absorption　吸湿性
57. permeability　渗透性
58. easy-wrinkling　易皱
59. easy-shrinking　易缩水
60. intensity　强度
61. heat-conduction　导热性
62. hand-feeling　手感
63. wearable　穿戴舒适
64. hard-storage　难储存
65. wrinkle resistance　抗皱性
66. static electricity　静电
67. hot melting/ fusibility　热熔/熔融性
68. elasticity　弹力
69. fade　掉色；褪色
70. sun resistance　耐晒性
71. acid and oxidant resistant　耐酸、耐氧化剂
72. drape　悬垂性
73. rigidity　挺扩性

Exercises

1. **Describing one of your garment's materials in English in video for 2−5 minutes.**
2. **Distinguish what is Twill & Drill & Chino?**
3. **Translating the foreign trade case into Chinese.**

Dear Nicole,

　　Request price quote for 6,000 each aprons , drill material , color khaki and 6,000 each bandanas same material , same color . This will require a logo and we will let you know later. The drill material here in Panama is similar to the one used for jeans , drill 60/20 or drill 65/36. The material must be something that will be washable and durable. 6,000 each TAN COLOR.

　　The size should be 87cm × 78 cm, full length. The picture you sent of the blue one first time looks like the one we need.

The bandana size is 19″ ×30″ and must be the same color (TAN) of the apron - no print on it, if the same material of the apron is not to rough and can be used to cover the head, then the same material used for the apron .6,000 each TAN color.

MANY TKS!

B.RGDS,

Gary

导入：在服装外贸中，T/R，R/C，T/V是什么意思？

第二章　面料

服装面料是指用来制作服装的材料。作为服装三要素之一，面料不仅可以诠释服装的风格和特性，而且直接左右着服装的色彩、造型的表现效果。

第一节　面料纤维

纺织纤维可分为天然纤维与人造纤维。天然纤维可进一步分为植物纤维、动物纤维和矿物质纤维。

"纺织"这一术语原本是用来描述机织布料的，现在，此术语已普遍应用在通过交织方式获得天然和人造纤维，以及长纤丝或纱线的任何制作上（交织程序是把经纱与纬纱通过交织方式形成布料的过程）。例如，通过机织、针织、制毯、毡合和起毛等方式制成的缝线、绳线、绳索、织带、花边、绣制品、网布和衣服等。

学习面料的第一步是明白纺织纤维分类。纤维可根据它的最初特性与生物类种进行详细的分类（图2-1）。

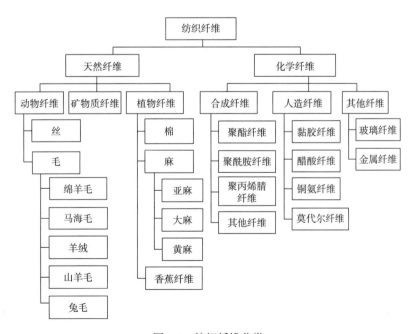

图2-1　纺织纤维分类

天然纤维有不同的性能，普遍价格偏高，其中有些价格非常昂贵。棉纤维是最重要的天然纤维，因为在世界纤维应用中它占半成。麻是另一种重要的天然纤维素纤维，但生产成本要比棉昂贵得多。最重要的动物纤维是羊毛、马海毛、羊绒等，可用于制作套装、大衣、长裙、针织服装、地毯等。另一种动物纤维是真丝纤维，通常比较昂贵，可用来做女式罩衣或衬衫。

相对于天然纤维，化学纤维价格比天然纤维低，应用广泛，但性能较差。聚酯纤维是应用量仅次于棉纤维的一种重要的化学纤维。在服装行业，尼龙的产生要早于聚酯纤维。腈纶纤维因价格便宜和可机洗的优点占有针织服装业的重要市场份额。人造纤维一般是将天然纤维素溶解后经过化学处理和机械加工生产出来的，因此也称"再生"纤维。人造纤维被广泛应用在纺织服装行业，用作里料、工作服、长裙甚至家居用品。

第二节　面料分类

对面料的分类，通常采用以下两种分类标准。

按照织物结构的标准分为机织/梭织面料❶、针织面料和非织造面料。机织面料和针织面料是服装面料最常见的两类。非织造面料是由经黏合、熔合或其他机械、化学方法加工而成。

机织面料是把经纱和纬纱相互垂直交织在一起形成的织物。经向的纱线称为经纱，纬向的纱线称为纬纱。机织面料一般分为平纹、斜纹、缎纹三种。机织面料的优点是结构稳定、不容易变形、布面平整、可对其用印花、提花处理。缺点是弹性差。常用机织面料主要有雪纺、牛津布、牛仔布、斜纹布、法兰绒、花缎等。机织面料常应用于外套、西装、牛仔裤、裤装等（图2-2）。

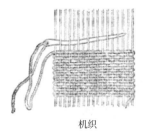
机织

纬纱

经纱

图2-2　机织面料的结构

针织面料因线圈是纱线在空间弯曲而形成的织物。每个线圈均由一根纱线组成，用织针将纱线或长丝钩成线圈，再把线圈相互串套形成线圈结构。针织面料分为经编和纬编。针织面料分为单面和双面。优点是伸缩性好、柔软性好、吸湿透气性好、抗皱性好。缺点是易脱散、尺寸稳定性差、易勾丝和起毛。常用的针织面料主要有汗布、天鹅绒、鸟眼布、网眼布

---

❶ 以下统称为机织面料。

等。针织面料通常应用于圆领衫、T恤衫、运动装、毛衫等（图2-3）。

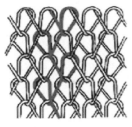 

经编　　　　　　　　　　　　　　　纬编

图2-3　针织面料的结构

按照纺织纤维成分的标准分为：天然面料、化学面料和混纺面料。

天然面料成本较高，化学面料成本低，但性能较低。因此，市场上面料大多都是混纺产品。混纺可以降低产品的成本，并提高产品的性能。

第三节　常用服装面料

在服装行业中，常用的服装面料主要有：棉、麻、丝、毛、皮革、涤纶、氨纶、锦纶、腈纶、莫代尔等。

1　棉

棉是以棉花为原料，经纺织工艺生产的面料。它的优点是轻柔贴身、保暖性好、吸湿性好、透气性好。它的缺点则是易缩水、易起皱、需要经常熨烫。它多用来制作时装、休闲装、内衣和衬衫。

2　麻

麻是从各种麻类植物取得的纤维制成的面料。一般有苎麻、亚麻、黄麻等。它的优点是强度高、导热好、吸湿好。缺点是硬、手感差。一般被用来制作休闲装、工作服，目前也多以其制作普通的夏装。

3　丝

丝属于蛋白纤维，是最富质感的天然纤维。它的优点是质地轻、柔软滑爽、高雅华丽、色泽鲜艳、吸湿性好。它的缺点是易生褶皱、褪色较快。它可被用来制作各种高档服装，尤其适合用来制作女士服装。

4　毛

毛属于蛋白质纤维，是各类羊毛、羊绒、兔毛、骆驼毛等织物的泛称。它的优点是穿戴舒适、手感柔软、高雅挺括、富有弹性和保暖性强。它的缺点主要是难洗涤。适用以制作礼服、西装、大衣等正规、高档的服装。

5　皮革

皮革是猪皮革、牛皮革、羊皮革、马皮革、驴皮革、袋鼠皮革、鳄鱼皮革、鸵鸟皮革等的泛称。它的优点是轻盈保暖，雍容华贵。它的缺点则是价格昂贵，对储存要求高。一般用于制作时装、冬装。

6 涤纶

涤纶是合成纤维中的一个重要品种，价格也相对便宜。涤纶的用途很广，大量用于制造服装面料和工业制品。它的优点是不易缩水、不易褪色、耐热性好、抗皱性好。它的缺点是吸湿性差、透气性差、易起静电、热熔/熔融性差。

7 氨纶

氨纶译名"斯潘德克斯"，是一种弹性纤维，缩写为PU。它的优点是弹性好，能够拉长6~7倍。缺点是吸湿性差。

8 锦纶

锦纶又叫尼龙，是世界上第一种合成纤维，它的出现使纺织品的面貌焕然一新，它的合成是合成纤维工业的重大突破。它的优点是不易缩水、不易褪色、抗皱性好、弹性好。它的缺点是透气性差、吸湿性差、耐热性差和易起静电。

9 腈纶

腈纶即聚丙烯腈纤维。腈纶的性能与羊毛非常相似，弹性较好，伸长20%时回弹率仍可保持在65%，蓬松、卷曲而柔软，保暖性比羊毛高15%，有合成羊毛之称。它的优点是弹性好、耐晒性好、保暖性好，耐酸、耐氧化剂。它的缺点是耐碱性差、吸湿性差、热熔性差。

10 莫代尔

莫代尔是一种纤维素纤维，是纯正的人造纤维。由产自欧洲的灌木制成木浆后，经过专门的纺丝工艺制作而成。纤维可与多种纤维混纺、交织，如棉、麻、丝等，多用于内衣生产。它的优点是柔软、光滑、色泽艳丽、悬垂性好、吸湿性好、色牢度好。它的缺点是挺扩性较差。

随着科技创新，菠萝、香蕉、苹果、咖啡等这些日常生活中最常见的食物，现在已可作为时尚行业的可持续的天然纺织原材料，并且得到广泛应用。

Basic Garment Knowledge
服装基础知识

Chapter 3　Accessories
第三章　辅料

项目名称：辅料

项目内容：1. 里料

2. 衬料

3. 垫料

4. 填料

5. 缝纫线

6. 扣紧材料

7. 其他辅料

教学学时：6课时

教学目的：让学生了解对辅料的基本概念、分类和特点，并培养学生服装辅料专业术语英文表达能力。

教学方式：由教师通过课前导入任务引出本章内容。结合案例分析讲解辅料分类、特点及其运用。观看拓展视频，了解几种基本缝纫线迹的区别。结合外贸案例，深入学习常用服装辅料的英文表述。总结归纳学习难点和重点，学生进行实操练习。

教学要求：（1）掌握本课词汇。

（2）了解辅料的分类及特点。

（3）熟练掌握常用服装辅料的英文表述。

Chapter 3　Accessories

In a garment, all the materials except fabrics are called accessories, which are the essential elements for function expansion and garment decoration. Depending on the function, accessories can be subdivided into the following seven categories.

Part 1　Lining

Lining is one of the earliest garment accessories used in garment, mainly including polyester taffeta, nylon, flannelette, all kinds of cotton and CVC/TC.

Function of lining are: make the garment smooth, convenient and comfortable; reduce the friction between fabric and underwear to protect fabric;increase the thickness of garment to keep warm; make garment flat and neat. As the lining of wadding clothing, it can prevent the wadding from being exposed; as the lining of jacket, it can keep the fur from getting dirty and keep it tidy.

According to their manufacturing process, lining may be classified into detached lining and fixed lining.

According to their fiber contents, lining may be classified into three groups: lining made from natural fiber, lining made from man-made fiber, and lining made from blended fiber.

Part 2　Interlining

Interlining, which plays a role in shaping or warmth, is used between the ordinary lining and the outside fabric. It is mainly used for front,shoulders, chest, collar, cuffs, pockets and waistband.

Interlining can improve the wearing comfort of garment. With the help of the stiffness and elasticity of the lining, it can shape, maintain the shape, support, flatten and reinforce the garment. Improve the wrinkle resistance and strength of garment. Improve the wearing performance and service life of garment, and improve the processing performance.

Based on the raw materials. interlining can be divided into four categories： cotton lining, linen lining, animal hair lining and chemical lining.

Part 3 Pad

Garment pads are used to support specific parts of garment to meet the design requirements, such as heightening, thickening, leveling and decoration, so as to give fullness and curve appearance.

Pads mainly includes chest pad, shoulder pad, collar pad and hip pad.

Chest pad mainly plays the role of comfortable wearing and improving the chest shape. Shoulder pad is shaped pad used for garment shoulders to support shoulder shape or lengthen shoulder line. The use of shoulder pad improves the quality of garment and has a luxurious feeling. Common collar pads include collar supporting and collar stay, which play the role of pursuing three-dimensional feeling, flatness and standardization.

Part 4 Filling

Filling is the soft material between the shell fabric and the lining. The main function is to keep warm and cool, radiation protection, health care and moisture absorption.

Garment fillings can be divided into cotton, down, silk cotton, natural fur/wool, fake fur/wool, camel down, space cotton, etc.

Part 5 Sewing Thread

Sewing thread is the material that must be used when stitching garment pieces. It is mainly used for sewing and decorative purposes, and it affects the quality, efficiency and appearance of the sewing procedure.

Sewing thread is mainly used for sewing, overlocking, topstitching and quilting, etc.

Sewing threads are often divided into staple sewing thread, filament sewing thread and combination sewing thread.

Part 6 Fasten Materials

Fasten materials are the materials used to fasten garment pieces, and also have a decorative role. There are mainly buttons, zippers, hooks, rings, buckle, string, velcro etc.

Button is one of the most commonly used fastening materials in garment.According to the material used, button can be divided into synthetic button, metal button, wood button, leather button, shell button, cloth button and so on. Commonly used buttons include: snap, tag button, snap button, snap fastener, etc.

In international trade, the specification unit of the button is Ligne❶.

As a very important fastening material in garment, according to the structure of zipper, it can be divided into open-end zipper and close-end zipper; according to the material of the teeth, it can be divided into nylon zipper，metal zipper, derlin zipper etc.

Part 7　Other Accessories

In addition to the above six commonly used main accessories, other accessories include belt materials, decoration materials, identification materials and packaging materials.

Band materials mainly include elastic, rib, waist tag, etc.Decorative materials mainly include decorative materials (beads, sequins, plastic sheets, animal feathers, etc.), lace, bows, badges etc. Identification materials mainly include trademarks, hangtags, string seal etc. Packaging materials mainly include wooden cases, cartons, inner boxes, paper bags, etc.

New Words and Expressions

1. accessorie　辅料
2. lining　里料
3. taffeta　塔夫绸
4. nylon　尼龙
5. flannelette　法兰绒
6. CVC/TC　涤棉
7. friction　摩擦
8. detached　可拆卸的
9. fixed　固定的
10. interlining　衬料
11. pad　垫
12. collar supporting　领托
13. collar stay　领插竹
14. radiation　辐射
15. space cotton　太空棉
16. down　羽绒
17. fake fur　人造毛皮
18. filling　填料
19. sewing thread　缝纫线
20. overlocking　拷边；锁边
21. topstitching　缉缝
22. quilting　绗缝
23. staple　短纤维
24. filament　长纤维
25. combination　混合
26. button　纽扣；扣子
27. zipper　拉链
28. hook　钩
29. ring　环
30. buckle　搭扣

---

❶ Ligne is always abbreviated to L.

31. string　绳带
32. synthetic　合成的
33. metal　金属的
34. shell　外壳
35. snap　按扣
36. tag button　工字扣
37. snap button　四合扣；大白扣
38. snap fastener　揿扣；手缝扣；子母扣
39. derlin　树脂
40. elastic　橡皮筋
41. rib　罗纹带
42. waist tag　腰卡

43. beads　珠子
44. sequin　亮片、金属片
45. plastic sheet　塑料片
46. animal feather　动物羽毛
47. bow　蝴蝶结
48. badge　肩章
49. trademark　商标
50. hangtag　吊牌
51. string seal　吊粒
52. wooden case　木箱
53. carton　纸箱
54. paper bag　纸袋

Exercises

Translate the accessories into Chinese in the order below.

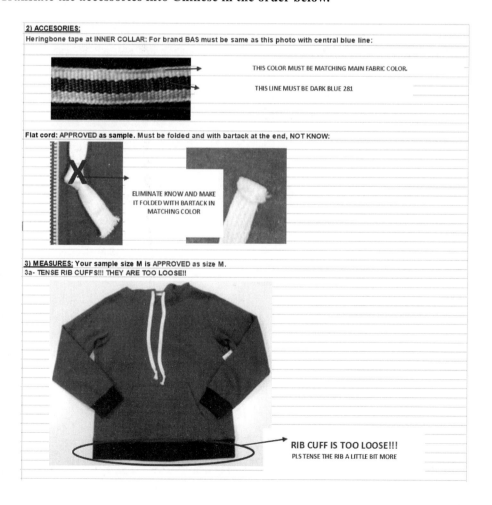

导入：在服装中，你知道哪些辅料？有什么功能？

第三章　辅料

在服装中，除了面料以外，用于服装上的一切材料都称为服装辅料，它是扩展服装功能和装饰服装必不可少的元素。根据功能不同，辅料可以细分为：里料、衬料、垫料、填料、缝纫线、扣紧材料和其他材料。

第一节　里料

里料是出现最早的一种服装辅料，用于服装夹里的材料，主要有塔夫绸、尼龙、法兰绒、各类棉布与涤棉布等。

里料的主要作用是：使服装穿脱滑爽方便，穿着舒适；减少面料与内衣之间的摩擦，起到保护面料的作用；增加服装的厚度，起到保暖的作用；使服装平整、挺括。提高服装档次；絮料服装的夹里，可以防止絮料外露；作为皮衣的夹里，它能够使毛皮不被沾污，保持毛皮的整洁。

服装里料种类多样。根据工艺分为分离式衬里和固定式衬里。

根据纤维成分，里料可分为三类：天然纤维里料、化学纤维里料和合成纤维里料。

第二节　衬料

衬料即衬布，是附在服装面料和里料之间的材料，起定型或保暖的作用。主要用于服装的前身、肩、胸、衣领、袖口、袋口、裤腰。

服装衬料可以提升服装的穿着舒适性。借助于衬的硬挺和弹性，可以对服装起塑型、保型、支撑、平挺和加固的作用。提高服装的抗皱能力和强度。提高服装的服用性能和使用寿命，并能改善加工性能。

根据原料不同，将衬料分为棉布衬、麻衬、动物毛衬和化学衬四大类。

第三节　垫料

垫料是在服装的特定部位，用垫物做支撑以达到设计的要求，例如加高、加厚、平整、修饰等，从而赋予服装丰满和曲线的外观。

垫料主要分为胸垫、肩垫、领垫和臀垫。

胸垫主要起到穿着舒适、改善胸型的作用。垫肩是经过定型的衬垫，用于服装肩部，起托起肩部形状或拉长肩线的作用。垫肩的使用提高了服装的质量，给人一种华贵的感觉。常见的领垫包括领托和领插竹，起到追求立体感、平整度、规范感的作用。

第四节　填料

服装填料，就是位于面料和里料之间起填充作用的材料。主要功能是保暖、降温、防辐射、保健、吸湿。

服装填料可分为棉花、羽绒、丝棉、天然毛皮/羊毛、人造毛皮/羊毛、驼绒、太空棉等。

第五节　缝纫线

缝纫线是缝制服装衣片时必须使用的线类材料。缝纫线主要起缝合、装饰作用，它会影响缝制的质量、效率和外观。

缝纫线主要用于缝合、拷边/锁边、缉缝、绗缝等。

缝纫线通常分为短纤维缝纫线、长丝缝纫线、混合缝纫线等。

第六节　扣紧材料

扣紧材料是服装中起封闭、扣紧作用的材料，也有一定的装饰作用。主要有纽扣、拉链、钩、环、搭扣、绳带、魔术贴等。

纽扣是服装中最常用的扣紧材料之一。按材质可分为合成材料纽扣、金属纽扣、木质纽扣、皮革纽扣、贝壳纽扣、衣料布纽扣等。常用的纽扣包括：按扣、工字扣、四合扣/大白扣、揿扣/手缝扣/子母扣。

在国际贸易中，常用的纽扣的规格单位是莱尼。

拉链作为服装中非常重要的扣紧材料，按拉链的结构分为开尾拉链和闭尾拉链；按拉链链齿的材料可分为尼龙拉链、金属拉链、树脂拉链等。

第七节　其他辅料

在服装中，除了以上六种常用的主要的辅料外，其他辅料还有带类材料、装饰材料、标识材料和包装材料。

带类材料主要包括松紧带、罗纹带、腰卡等。装饰材料主要有缀饰材料（珠子、亮片、塑料片、动物羽毛等）、花边、蝴蝶结、肩章等。标识材料主要有商标、吊牌、吊粒等。包装材料主要有木箱、纸箱、纸盒、纸袋等。

Basic Garment Knowledge
服装基础知识

Chapter 4　Color and Application in Fashion
第四章　色彩在服装中的应用

项目名称：色彩在服装中的应用

项目内容：1. 色彩基本知识

2. 常用色彩的英文表达

教学学时：2课时

教学目的：让学生了解服装中色彩的基本知识，以及常用色彩的英文表达。

教学方式：通过课前导入任务引出本章内容。分析讲解色彩的基本知识。学习常用色彩的英文表达。观看拓展视频，深入学习色彩程度的英文表述。总结归纳学习难点和重点，学生进行实操练习。

教学要求：（1）掌握本课词汇。

（2）了解服装中色彩的基本知识。

（3）熟练掌握常用色彩的英文表达。

Chapter 4
Color and Application in Fashion

Part 1   Basic Knowledge of Color

Color is the first impression of garment sense. As one of the three elements of garment , it has a strong attraction. If you want to give full play to it in garment, you must fully understand the characteristics of color.

1   Three Components of Color

Understanding hue, bright and saturation is critical for creating beautiful color harmonies. These are the basic three components of color.

Hue, also known as hue, refers to the appearance and name of each color, which depends on the main wavelength of the color. Such as red, orange, yellow, green, blue, purple, etc. Hue is the main basis for distinguishing colors and the biggest feature of colors. Bright is the lightness and darkness of color. The color is mixed with pure white to get a light color; Mix it with pure black to make it dark; Adding different amounts of black and white will get different depth effects. In neutral color, the color with the highest lightness is white and the color with the lowest lightness is black. In color, any pure color has its own lightness characteristics.Saturation, also known as purity, refers to the bright or bright degree of color. For the purity of color, the difference method is calculated according to the degree of grey in this color. The purity of color refers to adding different amounts of grey to the pure color. The more grey is added, the lower the purity of color is; The less grey is added, the higher the purity of the color. Thus, turbid colors with different purity can be obtained. We call these colors high purity color, medium purity color and low purity color.

2   The Color Wheel

In color theory, color harmony refers to aesthetically pleasing and harmonious color combinations based on geometric relationships on the color wheel. Back in the late 17th century, Sir Isaac Newton created a circular diagram of colors or color wheel that would be the base of color theory and a revolution in understanding the relationships between colors. That color wheel, which was made of the seven colors of the rainbow, was later improved to 12 different hues. The color wheel has three predominant patterns: primary colors, secondary colors, and tertiary colors.

Primary colors: The three primary colors are spaced equally around the color wheel, they are red, yellow, and blue. They can be mixed together to create any other color. Every color derives from a mixture of the three primary colors and no colors can be mixed to create them. Secondary colors are orange, green, and purple. These are created directly from a combination of two primary colors. You can think of them as children of the primary colors. They are spaced equally around the wheel but exactly centered between the primary colors. When primary colors and secondary colors are combined, they create a third class of colors called tertiary colors. Tertiary colors are yellow-orange, red-orange, red-purple, blue-purple, blue-green, and yellow-green.

3　Chromatic Color, Neutral Color and Metallic Color

Color can be divided into chromatic color, neutral color and metallic color.Chromatic color refers to all colors included in the visible spectrum. It takes red, orange, yellow, green, blue and purple as the basic colors. Tens of thousands of colors produced by different amounts of mixing between basic colors and between basic colors and non colors belong to the color system. Neutral color, is composed of black, white, gold, silver and grey. It plays a role of spacing and harmony in the collocation of colors. Metallic color refers to the color with metallic luster. Common are: gold (golden yellow), silver (silver white), copper (rose red), iron (silver grey), nickel (silver white), etc.

Part 2　Commonly Used English Expressions of Color

In foreign trade, In order to use color accurately and normatively, international standard color cards are generally used, such as Pantone color card❶. For globalization, color standardization plays a very important role, enabling different manufacturers and customers engaged in cross-border trade to accurately select colors. Let's look at the commonly used English expressions of color (Table 4-1).

Table 4-1　Commonly Used English Expressions of Color

No.	Red Series	No.	Yellow Series
1	vermeil	1	deep orange
2	pink;rose bloom	2	light orange
3	plum;crimson	3	lemon yellow
4	rose	4	maize
5	peach blossom	5	olive yellow

❶　PANTONE color card is a world-renowned color authority, covering the color communication system in printing, textile, plastic, drawing, digital technology and other fields, and has become the international unified standard language for color information exchange today.

Continued

No.	Red Series	No.	Yellow Series
6	cherry	6	straw yellow
7	salmon pink	7	mustard
8	garnet	8	broze yellow
9	purplish red	9	york yellow
10	ruby red	10	rattan yellow
11	agate red	11	nude
12	coral	12	sunny yellow
13	iron oxide red	13	earth yellow
14	rust red	14	sand yellow
15	brick red	15	golden yellow ; gold
No.	**Blue Series**	**No.**	**Green Series**
1	ingigo	1	pea green ; bean green
2	blue	2	olive green ; olive
3	sky blue ; azure	3	tea green
4	moon blue	4	onion green
5	ocean blue	5	apple green
6	sea blue	6	forest green
7	acid blue	7	moss green
8	ice-snow blue	8	grass green
9	peacock blue	9	agate green
10	sapphire ; jewelry blue	10	crystal
11	powder blue	11	jade green
12	purplish blue ; navy	12	mineral green
13	navy blue	13	spearmint ; viridis
14	hyacinth ; purplish blue	14	peacock green
15	ultramarine	15	green black ;jasper

New Words and Expressions

1. impression　印象
2. element　要素；元素
3. characteristic　特点
4. component　组成部分；成分
5. hue　色调
6. bright　名度
7. saturation　饱和度
8. wavelength　波长
9. neutral　中性的
10. purity　纯度
11. turbid　浑浊的；混乱的
12. wheel　轮子；齿轮
13. combination　结合；联合
14. circular　圆形的；环形的
15. diagram　简图；图解
16. revolution　革命；巨变

17. harmony　和谐；协调
18. predominant　主要的；主导的
19. primary　初级的；主要的
20. secondary　次要的；从属的
21. mixture　混合物；混合
22. tertiary　第三的；第三位的
23. chromatic　彩色的
24. metallic　金属色
25. spectrum　光谱
26. copper　铜
27. nickel　镍
28. accurately　精确地
29. normatively　规范地
30. globalization　全球化
31. cross-border　跨国的

Exercises

Translate this paragraph into Chinese.

Ember Glow EXP-X14

EXP-X14, as a new product constantly updated, adopts a translucent upper with a variety of bright colors, presenting a vibrant color scheme. Bright pink and orange sequins follow the heel panel and subtle inner layer, while the rest of the upper part is characterized by saturated blue. The retail price is $120, and the latest EXP-X14 is now arriving at retailers.

导入：当消费者在服装店或网上选购时，第一影响因素是什么？

第四章　色彩在服装中的应用

第一节　色彩基本知识

服装色彩是服装感观的第一印象，作为服装三要素之一，它有极强的吸引力，若想让其在着装上得到淋漓尽致地发挥，必须充分了解色彩的特性。

1　色彩三要素

理解色相、明度和纯度对于创造美丽和谐的色彩至关重要，这是颜色的三个基本组成部分。

色相也叫色调，是指每种色彩的相貌、名称，它取决于该颜色的主波长，如红、橙、黄、绿、蓝、紫等。色相是区分色彩的主要依据，是色彩的最大特征。明度是色彩的明暗程度。颜色和纯白混合得到浅色，和纯黑混合得到深色。加入不同量的黑、白色会得到不同的深浅效果。在无彩色中，明度最高的色为白色，明度最低的色为黑色。在有彩色中，任何一种纯度色都有自己的明度特征。纯度也叫作饱和度，是指颜色的鲜艳或鲜明的程度。对于有彩色的纯度的高低，区别方法是根据这种色中含灰色的程度来计算的。色彩的纯度强弱是指在纯色中加入不等量的灰色，加入的灰色越多，色彩的纯度越低；加入的灰色越少，色彩的纯度越高。从而可以得出不同纯度的浊色，我们称这些色为高纯度色、中纯度色、低纯度色。

2　色轮（色相环）

在色彩理论中，色彩和谐是指基于色轮上的几何关系形成的在审美上令人愉悦、和谐的色彩组合。早在17世纪末，艾萨克·牛顿爵士就创造了一个颜色的圆形图表或色轮，这将是颜色理论的基础，也是理解颜色之间关系的一场革命。这个由彩虹的七色组成的色轮后来被改进为12种不同的颜色。色轮包含了三种主要模式，即原色、间色和复色。

三原色（基色）：三原色是红色、黄色和蓝色,在色轮周围平均分布。次生色：橙、绿和紫，是用两种三原色直接调和出来的颜色。复合色：当三原色和次生色组合时，它们会产生第三种颜色，称为复合色（第三色）。复合色为黄橙、红橙、红紫、蓝紫、蓝绿和黄绿。

3　有彩色、无彩色和金属色

色彩分为有彩色、无彩色和金属色。有彩色指包括可见光谱中的全部颜色，它以红、橙、黄、绿、蓝、紫等为基本色。基本色之间不同量的混合、基本色与中性色之间不同量的混合所产生的千万种色彩都属于有彩色系。无彩色也称为中性色，由黑、白、金、银、灰色组成。在色彩的搭配中起间隔、调和的作用。金属色是指带有金属光泽的颜色。常见有：金（金黄色）、银（银白色）、铜（玫瑰红色）、铁（银灰色）、镍（银白色）等。

第二节　常用色彩的英文表达

在对外贸易中，为了能准确和规范地使用颜色，一般都会使用国际通用的标准色卡，例如：潘通（PANTONE）色卡❶。对于全球化来说，颜色的标准化扮演着相当重要的角色，可以使从事跨国贸易的不同制造商和客户准确地选择颜色。来看一下常用颜色的英文表达（表4-1）。

表4-1　常用色彩的英文表达

序号	红色系列	序号	黄色系列
1	朱红	1	深橘黄
2	粉红色	2	浅橘黄
3	梅红	3	柠檬黄
4	玫瑰红	4	玉米黄
5	桃红	5	橄榄黄
6	樱桃红	6	稻草黄
7	橘红色	7	芥末黄
8	石榴红	8	杏黄
9	枣红色	9	蛋黄
10	宝石红	10	藤黄
11	玛瑙红	11	象牙黄
12	珊瑚红	12	日光黄
13	铁红	13	土黄
14	铁锈红	14	砂黄
15	砖红	15	金黄

---

❶ 潘通色卡是享誉世界的色彩权威，涵盖印刷、纺织、塑胶、绘图、数码科技等领域的色彩沟通系统，已经成为当今交流色彩信息的国际统一标准语言。

续表

序号	蓝色系列	序号	绿色系列
1	靛青	1	豆绿
2	蓝色	2	橄榄绿
3	天蓝，蔚蓝	3	茶绿
4	月光蓝	4	葱绿
5	海洋蓝	5	苹果绿
6	海蓝	6	森林绿
7	湖蓝	7	苔藓绿
8	冰雪蓝	8	草地绿
9	孔雀蓝	9	灰湖绿
10	宝石蓝	10	水晶绿
11	粉末蓝	11	玉绿
12	藏蓝	12	石绿
13	海军蓝	13	松石绿
14	紫蓝	14	孔雀绿
15	青蓝	15	墨绿

本模块微课资源（扫描二维码观看视频）

1.1 Garment Foreign Trading Company

1.2 Foreign Trade Documentary

1.3 Sewing Machine

2.1 Clothing Materials and Classification

2.2 Woven Fabric & Knitted Fabric

2.3 Natural Fiber

2.4 Man-made Fiber

2.5 Harris Tweed

3.1 Lining

3.2 Interlining

3.3 Pad

3.4 Filling

3.5 Button

3.6 Zipper

4.1 English Expression of Colors

Module 2
Garment Foreign Trade Documentary Translation

模块二
服装外贸单据翻译

Garment Foreign Trade Documentary Translation
服装外贸单据翻译

Chapter 5　Garment Parts
第五章　服装部位名称

项目名称：服装部位名称

项目内容：1. 西装外套各部位的中文表述

2. 夹克各部位的中文表述

3. 裤子各部位的中文表述

4. 连衣裙各部位的中文表述

教学学时：6课时

教学目的：让学生理解并掌握服装各个部位的中文表达和英文翻译；培养学生在外贸实务中对服装专业术语的理解能力和表达能力。

教学方式：由教师通过课前导入任务引出本章内容。结合图例讲解服装部位的中英文表述。通过外贸案例，深入了解服装各个部位的表达。总结归纳学习难点和重点，组织学生进行实操练习。

教学要求：（1）掌握本课词汇。

（2）熟悉服装各个部位的中文表达和英文翻译。

（3）在外贸案例中熟练运用。

Chapter 5　Garment Parts

For the personnel engaged in garment foreign trade, understanding the Chinese expression and English translation of various parts of garment is the basis of garment merchandising, marketing, translation and other foreign trade work. The following is to understand the Chinese expression of various parts of garment through the finished garment of top and bottom. The finished coat mainly introduces the Chinese and English expressions of various parts of suit, coat and jacket; The finished product of lower garment mainly introduces the Chinese and English expressions of various parts of trousers and skirts (skirt and dress).

Part 1　English Expressions of Parts of Suit Coat

English expressions of parts of suit coat are shown in Table 5-1.

Table 5-1　English Expressions of Parts of Suit Coat

No.	Part Name	No.	Part Name
1	stand collar/collar stand	3	lapel point
2	top collar/collar fall	4	lapel

Continued

No.	Part Name	No.	Part Name
5	mock hole	17	front underarm dart
6	breast pkt	18	side seam
7	fold line for lapel	19	armhole
8	button	20	sleeve top/sleeve head/crown
9	button hole	21	back lining
10	flap	22	hanging loop
11	front edge	23	across back shoulder
12	top fly/left front	24	back yoke
13	hem	25	top sleeve
14	under fly/right front	26	under sleeve
15	sleeve opening	27	center back seam
16	sleeve button	28	vent

Part 2   English Expressions of Parts of Jacket

English expressions of parts of jacket are shown in Table 5-2.

Table 5-2   English Expressions of Parts of Jacket

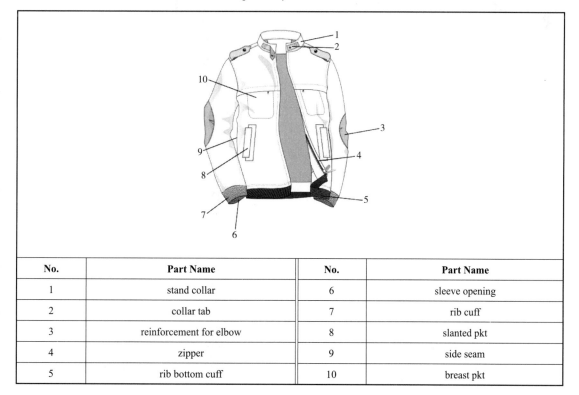

No.	Part Name	No.	Part Name
1	stand collar	6	sleeve opening
2	collar tab	7	rib cuff
3	reinforcement for elbow	8	slanted pkt
4	zipper	9	side seam
5	rib bottom cuff	10	breast pkt

Part 3   English Expressions of Parts of Pants

English expressions of parts of pants are shown in Table 5-3.

Table 5-3   English Expressions of Parts of Pants

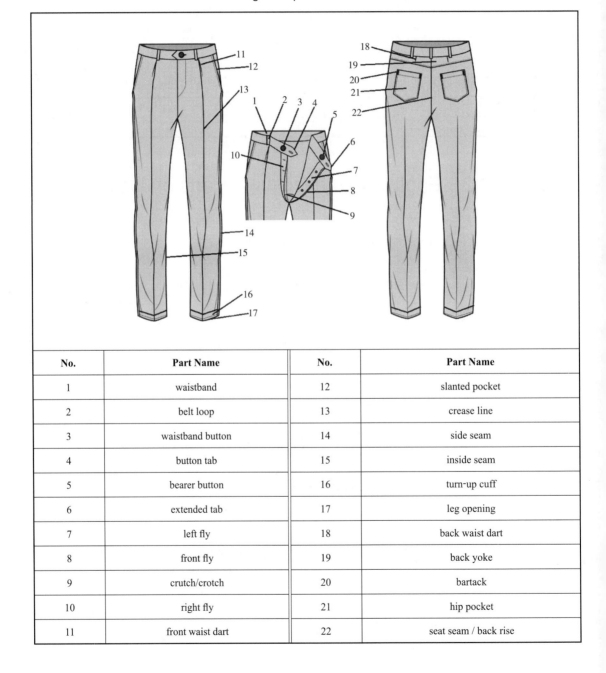

No.	Part Name	No.	Part Name
1	waistband	12	slanted pocket
2	belt loop	13	crease line
3	waistband button	14	side seam
4	button tab	15	inside seam
5	bearer button	16	turn-up cuff
6	extended tab	17	leg opening
7	left fly	18	back waist dart
8	front fly	19	back yoke
9	crutch/crotch	20	bartack
10	right fly	21	hip pocket
11	front waist dart	22	seat seam / back rise

Part 4 English Expressions of Parts of Dress

English expressions of parts of dress are shown in Table 5-4.

Table 5-4 English Expressions of Parts of Dress

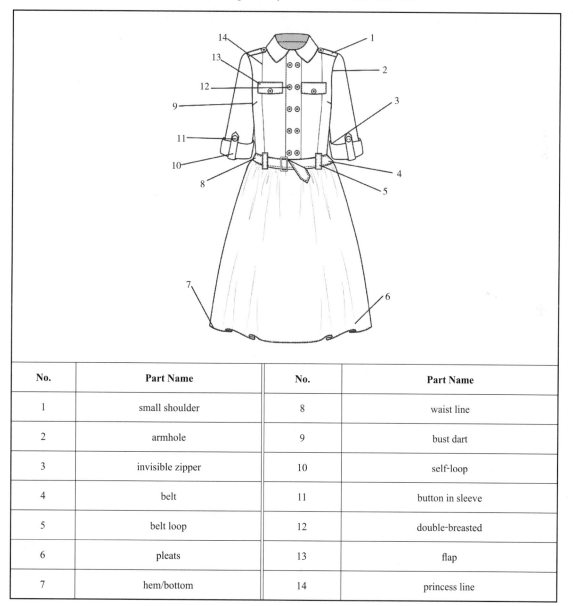

No.	Part Name	No.	Part Name
1	small shoulder	8	waist line
2	armhole	9	bust dart
3	invisible zipper	10	self-loop
4	belt	11	button in sleeve
5	belt loop	12	double-breasted
6	pleats	13	flap
7	hem/bottom	14	princess line

New Words and Expressions

1. stand collar/collar stand 领座；下级领
2. top collar/collar fall 领面；上级领
3. Mao collar 立领
4. collar tip 领尖
5. collar spread 领长
6. collar tab 领襻
7. collar stay 领插竹
8. collar notch 领子扣口
9. collar supporting 领托
10. self-loop 原身布带襻
11. armhole 袖窿
12. top fly/left front 门襟；左前幅
13. under fly/right front 里襟；右前幅
14. front dart 褶
15. sleeve opening 袖口
16. hem/bottom 底边
17. hem clean 净边
18. front edge 门襟止口
19. flap 袋盖
20. slanted pocket 斜插袋
21. breast pocket 胸袋
22. hip pocket 后袋
23. patch pocket 贴袋
24. lapel 驳头
25. fold line for lapel 翻领线
26. gorge line 串口线
27. across back shoulder 总肩；肩长
28. vent 背衩
29. yoke 育克；担干
30. side seam 侧缝；侧骨
31. waistband 腰头
32. button tab 里襟尖嘴
33. bearer button 腰头纽扣；承重钮
34. belt loop 串带
35. extended tab 宝剑头；裤头搭嘴
36. left fly 纽牌
37. crutch 裤裆；十字裆
38. inside seam 下裆缝
39. leg opening 裤脚
40. turn-up cuff 裤卷脚；反脚
41. crease line 烫迹线
42. front waist pleat 前腰褶
43. seat seam / back rise 后裆缝；后裆弧长
44. small shoulder 小肩
45. button hole 扣眼；纽门
46. bust dart 胸褶
47. invisible zipper 隐形拉链
48. belt 腰带
49. waist line 腰线
50. double-breasted 双排扣
51. bartack 打枣；套结
52. reinforcement for elbow 加固绸
53. rib cuff 罗纹袖口克夫；介英
54. back lining 后里布

Exercises

List the English expression of each part of the garments in the pictures below.

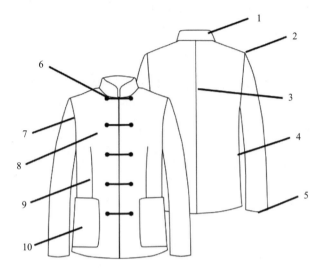

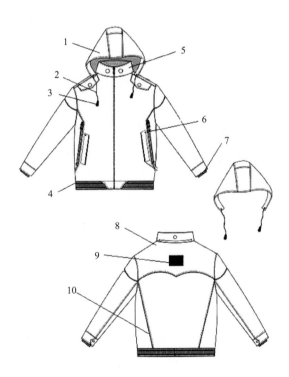

导入：在服装外贸中，如何用英文表述服装各个部位？

第五章　服装部位名称

对于从事服装外贸工作的人员来说，认识服装各个部位的中文表述和英文翻译是做好服装跟单、市场营销、翻译等外贸工作的基础。下面通过上装和下装的服装成品来认识服装各个部位的中文表述。上装的成品主要介绍西装、外套、夹克各部位的中英文表述；下装的成品主要介绍裤装和裙装（半身裙和连衣裙）各部位的中英文表述。

第一节　西装外套各部位的中文表述

西装外套各部位的中文表述见表5-1。

表5-1　西装外套各部位的中文表述

序号	部位名称	序号	部位名称
1	领座/下级领	4	驳头
2	领面/上级领	5	假扣眼
3	领嘴	6	胸袋

续表

序号	部位名称	序号	部位名称
7	翻领线	18	侧缝/侧骨
8	扣子	19	袖窿
9	扣眼	20	袖山
10	袋盖	21	后里布
11	门襟止口	22	挂耳
12	门襟/左前幅	23	总肩/肩长
13	衫脚	24	后育克/后担干
14	里襟/右前幅	25	大袖
15	袖口	26	小袖
16	袖扣	27	后中缝/后中骨
17	前肋褶	28	背衩

第二节　夹克各部位的中文表述

夹克各部位的中文表述见表5-2。

表5-2　夹克各部位的中文表述

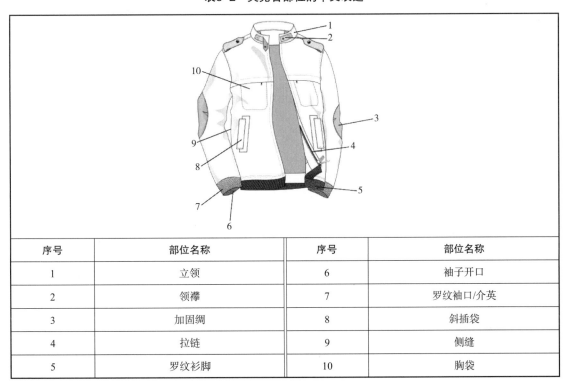

序号	部位名称	序号	部位名称
1	立领	6	袖子开口
2	领襻	7	罗纹袖口/介英
3	加固绸	8	斜插袋
4	拉链	9	侧缝
5	罗纹衫脚	10	胸袋

第三节　裤子各部位的中文表述

裤子各部位的中文表述见表5-3。

表5-3　裤子各部位的中文表述

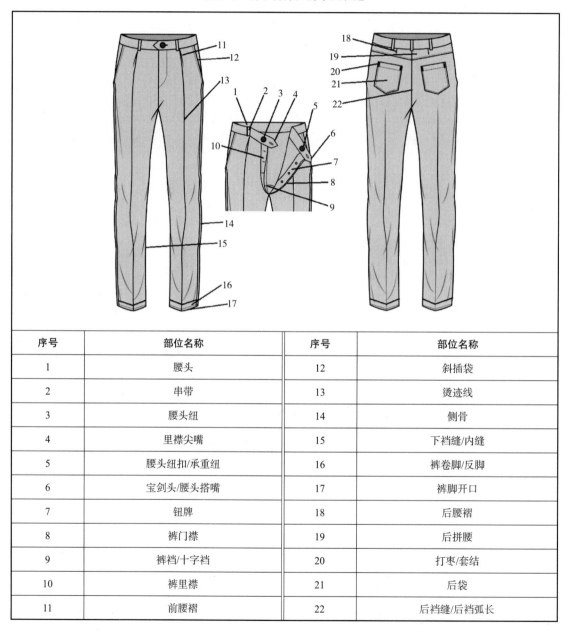

序号	部位名称	序号	部位名称
1	腰头	12	斜插袋
2	串带	13	烫迹线
3	腰头纽	14	侧骨
4	里襟尖嘴	15	下裆缝/内缝
5	腰头纽扣/承重纽	16	裤卷脚/反脚
6	宝剑头/腰头搭嘴	17	裤脚开口
7	钮牌	18	后腰褶
8	裤门襟	19	后拼腰
9	裤裆/十字裆	20	打枣/套结
10	裤里襟	21	后袋
11	前腰褶	22	后裆缝/后裆弧长

第四节　连衣裙各部位的中文表述

连衣裙各部位的中文表述见表5-4。

表5-4　连衣裙各部位的中文表述

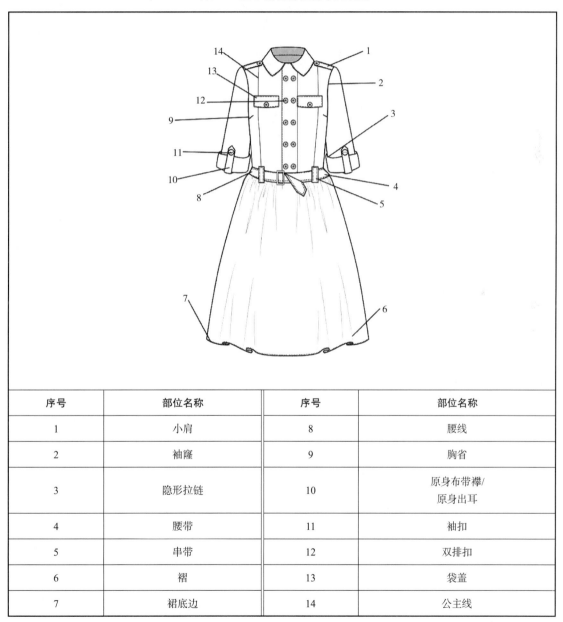

序号	部位名称	序号	部位名称
1	小肩	8	腰线
2	袖窿	9	胸省
3	隐形拉链	10	原身布带襻/原身出耳
4	腰带	11	袖扣
5	串带	12	双排扣
6	褶	13	袋盖
7	裙底边	14	公主线

Garment Foreign Trade Documentary Translation
服装外贸单据翻译

Chapter 6　Samples in Garment
第六章　服装样衣

项目名称： 服装样衣

项目内容： 1. 样衣的分类

2．服装的尺码及标注

教学学时： 4课时

教学目的： 让学生认识和掌握各类服装样衣的英文表述，了解其在服装外贸中的作用。

教学方式： 通过课前任务，引入本章内容；分析讲解外贸服装中样衣的种类和用途，以及尺码符号的区分；通过拓展外贸案例，让学生深入了解外贸服装。总结和归纳学习难点和重点，通过实操练习使学生进一步熟悉和掌握本章节内容。

教学要求： （1）熟悉掌握样衣的种类和用途。

（2）熟悉样衣的尺码。

（3）掌握各类样衣的英文表述。

Chapter 6　Samples in Garment

The "sample" in garment refers to the actual sample of garment. It can be divided into new samples designed and supported by enterprises and samples made according to customers and confirmed by customers.

The main functions of sample garment are:

(1) Evaluation of plant productivity.

(2) Confirm the color and color difference range.

(3) Check whether the fabric of bulk production is consistent with the sample.

(4) Production line and finished product inspection.

Part 1　Classification of Sample

1　Initial sample/proto sample

Initial sample is the sample made according to the design drawing, or the style that the factory gives its own customer's products to another customer for reference. It generally refers to the sample made to customers for the first time. The initial sample mainly focuses on style and technology. Fabrics and colors can be different.

2　Fitting sample

Fitting sample mainly depends on the effect and size of the model. For example: the balance of various parts (including up and down, front and rear, left and right) worn on the dummy. If the test is not good, must check the paper pattern.

3　Salesman/showroom sample

Sales samples are mainly used for exhibition samples of trade fairs or samples made for trial sale to customers before production. The purpose is to obtain orders through the exhibition samples for the new models made before. General sales samples are required to be of the same color and size, focusing on style and sales appearance.

4　Photo sample

Photo samples are provided to customers as models for trial photography. Some of them are to be used in printed books for reference in making color drawings of large goods. The main thing is that the color and style should be correct.

5   Pp(pre-production) sample

Pp sample is the sample before bulk production, which is mainly used to let the customers confirm bulk production. This sample is a complete reference sample for bulk production. Every details must meet the requirements, otherwise bulk production may also be wrong. If it does not meet the requirements, the customer may ask to do it several times until it meets the requirements.

6   Test sample

Test sample mainly tests whether the washing water, color and environmental protection meet the requirements of guests. Some customers require testing, so they need to do this.

7   Shipping sample/shipment sample

Shipping sample is mainly to confirm the quality of the goods to be delivered to the customers. Shipping sample represents the quality of bulk production, so the process should be consistent with bulk production and the workmanship should be better.

8   Counter sample

Counter sample is made by the factory according to the proto sample provided by the customer and sent back to the customer for confirmation. It is mainly to confirm the style, process and size for customers.

9   Bulk sample

Bulk sample is made of materials in bulk production, mainly to check the process, size, shrinkage and other problems. Bulk sample is not required by the customers, and there is no need to approve it for the customers, but the factory tries to make it for bulk production.

Part 2   Size Signs in Garment

In foreign trade activities, S, M, L, XL and XXL are the most commonly used sizes for garment in North America, two kinds of numbers are commonly used to represent the size: the first is to use 0, 2, 4, 6... the second is to use 34, 36, 38...

In China, 160 / 76A is commonly used to represent the model of garment. 160 is height, 76 is bust, and A is body shape, that is, the difference between chest and waist.The letters Y, A, B and C represent different body types. The "number" in garment size refers to the centimeter of a person's height, which affects the length of garment, sleeves and trousers; The "type" in garment size refers to the number of centimeters of a person's body circumference, and the parts it affects are chest circumference, waist circumference and hip circumference. For example, if your height is 170cm, your basic bust is 88cm and your basic waist is 73cm, you are suitable for wearing 170 / 88 tops and 170 / 73 pants.

Children's wear in foreign trade has special sizes, which are generally divided into three types:

(1) Mark according to the appropriate height: for example, 90cm, 100cm, 110cm, 120cm...

refers to the appropriate height of children.

(2) Mark according to the appropriate age: for example, 6M, 12M and 18M refer to the corresponding baby month, which is suitable for babies aged 6, 12 and 18 months; 4T, 5T, 6T (or 4A, 5A, 6A) refers to the corresponding children's age, which is suitable for children aged 4A, 5A and 6A.

(3) Distinguishing No. : for example, No. 2, 4, 6, 8, 10 or 00, 0, 1, 2, Just to distinguish the size of this dress, it does not necessarily represent age or height (Table 6-1).

Table 6-1　Expression of Common Garment Size Symbols

No.	Symbol	Meaning	No.	Symbol	Meaning
1	S（Small）	小码	9	0	小码
2	M（Medium）	中码	10	2	中码
3	L（Large）	大码	11	4	大码
4	XL（Extra Large）	加大码	12	6	加大码
5	XXL（Extra Extra Large）	加加大码	13	Y	胸围大腰细的体型
6	P or R（Petite/Regular）	比正常码数偏短	14	A	一般体型
7	L or T（Long/Tall）	比正常码数偏长	15	B	微胖型
8	W（Wide）	比正常码数偏宽	16	C	胖体型

New Words and Expressions

1. sample　样衣；样板
2. homonymous　同名的
3. initial/proto sample　原样；头样；初样
4. fitting sample　试身样
5. salesman sample　销售样
6. photo sample　照相样；影像样
7. pp sample/pre-production sample　大货样；产前样
8. shipping /shipment sample　船头样；船样
9. test sample　测试样
10. counter sample　对等样品；回样
11. bulk sample　先行样；货前样
12. bulk production　大货生产；批量生产
13. centimeter　厘米
14. circumference　围度
15. top　上衣
16. pant　裤子
17. petite　纤弱的；娇小的

Exercises

List the functions of different samples in Garment.

第六章　服装样衣

样衣是指服装的实际样品。可以分为企业自主设计出资制作的新款样品和按照客户要制作并经客户确认的样品。

样衣的主要作用有：

（1）对工厂生产力的评估。

（2）确认颜色以及色差范围。

（3）检查大货面料是否和样衣一致。

（4）生产线与成品检验。

第一节　样衣的分类

1　原样、头样、初样

原样是根据设计图做的样板，或者工厂把自己客户的产品给另外一个客户参考的款式。泛指第一次给顾客做的样板。原样主要是注重款式和工艺。面料、颜色都可以不同。

2　试身样

试身样主要是看模特穿起的效果以及尺寸问题。例如，穿在人体模型上各部位（包括上下，前后，左右）的平衡。如果试身不好，须查纸样。

3　销售样/展示室样品

销售样主要用于展销会的展样，或开货前做给客人试卖的样品，目的是将之前所做出的新款式，通过展销样品获得订单。一般销售样品要求齐色齐码，重在款式和销售外观。

4　照相样、影像样

照相样是提供给客人作模特试身照相的样板，有些是要作印刷用的，用于制作大货的彩图参考。主要是颜色和款式要正确。

5　大货样，产前样

大货样就是批量生产前的样衣，主要是让客人确认大货开货用的。此样品是做大货的完整参照样，所有细节都要符合要求，否则批量生产也可能会错。如果不符合要求，客人可能会要求做几次，直到符合要求为止。

6　测试样

测试样主要测试洗水、颜色、环保方面是否符合客人要求。有些客户要求做测试的，就需要做此样。

7 船头样、船样

船头样主要是给客人确定所要出的货品质量。船头样代表整批大货的质量，所以工艺要和大货一致，做工要更好。

8 对等样品、回样

回样是由工厂根据客户提供的原样制作并寄回客户经客户确认的样品。主要是确认款式、工艺、尺寸。

9 先行样、货前样

先行样采用大货布料来制作，主要是查看工艺、尺寸、收缩率等问题。先行样不是客人要求做的，也不需要客人批准，而是工厂自己为大货试做的。

第二节 服装的尺码及标注

在外贸活动中，服装尺码的标注最常使用的是S、M、L、XL和XXL。在北美洲，常用两种数字标注码数：第一种是使用0、2、4、6……第二种是使用34、36、38……

在中国，常用160/76A表示衣服的型号，160是身高，76是胸围，A是体型，即胸腰差。字母Y、A、B和C代表不同体型。服装号型中的"号"，是指人身高的厘米数，它影响的部位是衣长、袖长、裤长；服装号型中的"型"，是指人的体围厘米数，它影响的部位是胸围、腰围、臀围。例如，身高170cm，基本胸围为88cm，基本腰围为73cm，那么就适合穿170/88的上衣，170/73的裤子。

外贸童装标注尺码比较特殊，一般分三种：

（1）按适合的身高标注：例如，90cm、100cm、110cm、120cm……指的是适合的儿童身高。

（2）按适合的年龄标注：例如，6M、12M、18M，指的是相应的宝宝月份，适合6岁、12岁、18个月的宝宝穿着；4T、5T、6T（或是4A、5A、6A）指的是相应的儿童年龄，适合4岁、5岁、6岁的孩子穿着。

（3）区分编号：例如2，4，6，8，10号或是00，0，1，2号，只是为了区分此款衣服的大小号，不一定代表年龄或是身高（表6-1）。

表6-1 常用衣服尺码符号的表达

序号	符号	含义	序号	符号	含义
1	S（Small）	小码	9	0	小码
2	M（Medium）	中码	10	2	中码
3	L（Large）	大码	11	4	大码
4	XL（Extra Large）	加大码	12	6	加大码
5	XXL（Extra extra Large）	加加大码	13	Y	胸围大腰细的体型
6	P或R（Petite/Regular）	比正常码数偏短	14	A	一般体型
7	L或T（Long/Tall）	比正常码数偏长	15	B	微胖型
8	W（Wide）	比正常码数偏宽	16	C	胖体型

Garment Foreign Trade Documentary Translation
服装外贸单据翻译

Chapter 7　Sample Order and Approval Sample Card
第七章　样板单和确认样板卡

项目名称：样板单和确认样板卡

项目内容：1．样板单

　　　　　　2．确认样板卡

教学学时：6课时

教学目的：让学生了解成衣生产中样板单的作用和格式；能熟练翻译英文样板单和确认样板卡，培养学生服装外贸相关工作的熟练翻译能力。

教学方式：由教师通过课前导入任务引出本章内容。分析讲解成衣生产中样板单的作用、格式和翻译。通过拓展外贸案例，强化样板单的翻译部分。归纳本章学习难点和重点。组织学生进行实操练习。

教学要求：（1）掌握本课词汇。

　　　　　　（2）熟练掌握样板单和确认样板卡中的专业术语。

　　　　　　（3）了解制作样板单的格式和要点。

　　　　　　（4）能阅读和翻译英文样板单和确认样板卡。

Chapter 7
Sample Order and Approval Sample Card

Part 1   Sample Order

Sample is sent to customer for approval. Sample order is made out by Sales Department (Table 7-1). The specification of sample size and raw material are listed on sample order. Four copies are used with the following distribution: one to Sales Department, one to Sample Room, two to Production Department (i.e.pattern maker and cutting supervisor). For example:

Table 7-1   Sample Order

SAMPLE ORDER			
TO: _____ DATE: _____ DELIVERY DATE: _____			
CLIENT: _____ STYLE NO.: _____ OUR REF NO.: _____			
PATTERN NO: _____ DISCIPTION: _____ LABEL: _____			
SHELL FABRIC: _____ LINING: _____ SIZE: _____ QUANTITY: _____			
Description	**Requested Meas.**	**Finished Meas.**	**Sketching**
1/2 Waist Width-Relaxed			
1/2 Waist Width-Stretched			
Waist Length			
Front Rise			
Back Rise			
Side Seam			
Inner Seam			
1/2 Bottom Width (vertical measure)			
Pocket			
Front Length			
Center Back Length			
Shoulder Width			
1/2 Chest Width (2cm under armhole)			
1/2 Hips/Seat			

Continued

Description	Requested Meas.	Finished Meas.	Sketching
Collar High			
1/2 Armhole（curve straight）			
Sleeve Length（including cuff）			
Sleeve Opening			
1/2 Cuff Width			
Cuff Length			
Others			
Fabric Sample：		Accessories Sample：	

Materials Request:

1.#604 top thread/ #604 under thread/ T.T.T. thread, T.T.T.#403 overlock, #402 bartack.

2.#24L black /rainbow agoya button, applied in back pocket.（Customer supply）

3. Collar, cuff, pocket, T.T.T. satin with tapes.

4. T.T.T. lining #190.

5. Size label/ yard label. Hanging in center back collar.

Stitch Density:Lockstitch machine/double needles, no less than 10 stitches/inch, overlock stitch machine（third thread-no less than 11 stitches/inch, fifth thread-no less than than 12 stitches/inch）

Sewing: Smooth, even comfortable, no open stitch, broken stitch, skipped stitch, overlapped stitch.Quality assurance, front edge with bartack. No reworking.

Ironing: Keep cloth clean and workshop orderly. Well pressed without dirty, over-pressed, iron-shinning, waterspot, color-shading.

Remarks:

Signature of Merchandiser: _____

Signature of Patronier: _____

Signature of Sample Supervisor: _____

Part 2　Approval Sample Card

That the sample is approved for bulk production is subject to the instructions and comments on the approval sample card. This is a very important to confirm all the details are corrected. It is made out by Sales Department. The information must contain:

(1) Client.

(2) Factory.

(3) Style number.

(4) Fabric.

(5) Contract number.

(6) Size specification and measurement.

(7) Color.

(8) Garment constructions.

(9) Trimmings.

Five copies are made under the following distribution: one is kept by Sales Department, four to Production Department for grading, cutting, sewing and quality controlling (Table 7-2).

Table 7-2　Approval Sample Card

APPROVAL SAMPLE CARD				
Client:　　　　　　Factory:　　　　　　Style No.:				
Shell Fabric:　　　　Contract No.:　　　　Size Spec.:				
Dept.	**Style**	**Q.C.**	**Merchandising**	**Date**

The attached sample is approved for bulk production subject to the instructions and comments on this card (Table 7-3).

Table 7-3　Approval for Bulk Production

Size:

Measure Point	**Spec.**	**Actual**	**Measure Point**	**Spec.**	**Actual**
Waist			Waist		
Hips/Seat			Bottom		
Front Rise			C.B. Length		
Back Rise			Shoulder		
Thigh			Armholes		
Knee			Sleeve Width		
Bottom			Sleeve Opening		
Inseam			Sleeve Length		
Zipper			Collar		
Out Seam			Neck Drop		
Chest			Neck Opening		

Following points must be corrected in production:

_____

Approval before production:

Front Pocket _____ Back Pocket _____

Trimmings _____

Others _____

Continued

GARMENT CONSTRUCTIONS

1)	Size	Dimension			
Front Pocket	(　)	×　×			
	(　)	×　×			
	(　)	×　×			

2)	Size	Dimension			
Back Pocket	(　)	×　×			
	(　)	×　×			
	(　)	×　×			

3) Back Yoke Size _____ (at seat seam) _____ (at side seam)

4) Back Pocket Placement

　(a) _____ Below + Parallel to back yoke edge

　(b) _____ Below + Parallel to bottom of waistband

　(c) Bellow bottom of waistband _____ (at seat seam)

　　　　　　　　　　　　　　　　　 _____ (at side seam)

COLOR CO-ORDINATION

Body	Top-stitching	Emb.	Label	Metalware
_____	_____	_____	_____	_____
_____	_____	_____	_____	_____

SHELL FABRIC

Name/Code _____ Fabric Content _____

Weight _____ oz. per sq. ft

　　_____ lb. per doz.

Construction _____ × _____

　　　　　　　 S × 　　　　　 S

TRIMMINGS

	Accepted	Unaccepted
Top stitch thread size	(　)	(　)
Button/Snap	(　)	(　)
Rivet	(　)	(　)
Zipper	(　)	(　)
Interlining	(　)	(　)
Lining	(　)	(　)
Piping/Insert	(　)	(　)
Ribbing	(　)	(　)
Drawstring	(　)	(　)
Shoulder Pad	(　)	(　)

New Words and Expressions

1. distribution　分配；分布
2. client　顾客；客户
3. relax　使松弛；放松
4. stretch　伸展；张开
5. front rise　前裆弧线（前浪）
6. armhole　袖窿（夹圈）
7. accessories　辅料；配料
8. delivery date　交货日期
9. grading　推板；放码
10. Dept.（Department）　部门
11. Q.C.（Quality Control）　品质控制
12. Spec.（Specification）　规格
13. thigh　大腿围
14. neck drop　领深
15. dimension　尺寸；规格
16. yoke　育克
17. Emb.（Embellishment）　装饰
18. snap　揿纽；按扣；钩扣
19. rivet　铆钉
20. interlining　衬料
21. piping　（装饰衣服边的）管状窄条
22. insert　被嵌入之物（如衣裙上的花边等）
23. ribbing　针织罗纹镶边（用于领口、袖口、腰部等）
24. drawstring　束带；抽带
25. shoulder pad　肩垫

Exercises

1. List the difference functions of "sample order" and "Approval sample order".

2. Translate the e-mail about sample below.

Dear Nicole,

Attention to the PP SAMPLE submitted before August 15, 2020.

S. and B. opening w/ RIB cuff in MATCHING COLOR.Include herringbone tape in MATCHING COLOR at inner B.C.There is two darts in the B.P.The shell fabric must be CVC 65/35.

PLS SEND PP SAMPLE IN BULK FABRIC W/ CORRECT TRIMS FOR FINAL CONFIRMATION BEFORE BULK.PLS FOLLOW SAME R.AS O.S.MUST BE MELANGE EFFECT IF MAIN FABRIC IS MELANGE.

WORKMANSHIP: Very Good

MEASUERS:

Across Shoulder +2 cm is too much (Tolerance is 1cm)This is a problem.

B.O.on Fabric Edge + 4 cm is too much (Tolerance is 1cm) This is a problem.

B.O. on Rib Edge + 4 cm is too much (Tolerance is 1 cm)This is a problem.

N.W.: Should be 24 and has 26 cm . Two centimeters is a lot , please ask to be in tolerance of 1cm .

Hood L.: should be 41cm and has 43 cm . If possible make according to meas.

MANY TKS!

B.RGDS,

Gary

导入：如何理解和翻译外贸中的服装样板单和确认样板卡？

第七章　样板单和确认样板卡

第一节　样板单

样板须送给客户认可。样板单由销售部门填写。样板尺寸的规格和原材料列在样板单上（表7-1）。复印四份做如下分派：一份给销售部门，一份给板房，两份给生产部门（即唛架制作员和裁剪主管）。

表7-1　样板单

样　板　单			
送至：＿＿＿＿＿　发单日期：＿＿＿＿＿　交付日期：＿＿＿＿＿			
客户：＿＿＿＿＿　款式编号：＿＿＿＿＿　厂内编号：＿＿＿＿＿			
纸样编号：＿＿＿＿＿　品名：＿＿＿＿＿　商标：＿＿＿＿＿			
面料：＿＿＿＿＿　里料：＿＿＿＿＿　尺码：＿＿＿＿＿　数量：＿＿＿＿＿			
部位	需要尺寸	实际尺寸	款式图
1/2腰围（放松量）			
1/2腰围（拉伸量）			
腰节			
前裆（即前浪）			
后裆（即后浪）			
侧缝长			
内缝长			
1/2底边（直度）			
口袋			
前长			
后中长			
肩宽			
1/2胸围（袖窿下2cm度量）			

续表

部位	需要尺寸	实际尺寸	款式图
1/2臀围/坐围（肩点下60cm度量）			
领高			
1/2袖窿弧长（直度）			
袖长（含袖克夫）			
袖口宽			
1/2袖克夫宽			
袖克夫长			
其他			

面料样板：	辅料样板：		

物料要求：1. 604号面线／604号底线／配色线，配色403号线锁边，402号线打套结（打枣）。

2. 纽扣24L贝壳纽黑色／彩虹色。用于后袋（客供）。

3. 领边、袖口、袋口边配色缎纹布车夹条。

4. 里料用配色190号。

5. 织唛。吊车于后领中

针迹密度：平车／双针-不少于10针/英寸；锁边机（三线锁边机-不少于11针/英寸，五线锁边机-不少于12针/英寸）

车缝要求：各部位线路顺直、宽窄一致，不接受开线、断线、跳针、明显位置驳线。要求每道工序保证车缝质量、止口位必须打回针。请小心车缝，避免返工

整烫要求：请注意及时更换洁净烫布，做好现场保洁。要求各部位整烫平服，不允许弄污、烫黄、极光、水渍、变色等质量问题存在

备注：

跟单员签名：_____

纸样师签名：_____

板房主管签名：_____

第二节　确认样板卡

样板卡被确认为生产指令和说明是进行批量生产的前提条件。确认样板卡上的所有资料都必须是正确的，这一点很重要。确认板卡由销售部门填写。内容包含：

（1）客户。

（2）工厂。

（3）款式编号。

（4）面料。

（5）合同编号。

（6）规格尺寸。

（7）颜色。

（8）服装款式结构。

（9）辅料。

复印五份做如下分派：一份保存在销售部门，四份送给生产部门作为放码、裁剪、缝制和品质控制等（表7-2）。

表7-2　确认样板卡

确认样板卡				
客户：＿＿＿＿＿＿　　工厂：＿＿＿＿＿＿　　款式编号：＿＿＿＿＿＿				
面料：＿＿＿＿＿＿　　合同编号：＿＿＿＿＿＿　　尺码：＿＿＿＿＿＿				
部门	款式	质量控制员	货品	日期

所附样板卡经确认，可作为大货生产指令，并在此卡中说明（表7-3）。

表7-3　大货确认样板卡

			样板尺码：＿＿＿＿＿＿		
测量部位	规定尺寸	实际尺寸	测量部位	规定尺寸	实际尺寸
腰围			腰围		
臀围			下摆		
前裆弧长（前浪）			后中长		
后裆弧长（后浪）			肩宽		
大腿围			袖窿		
膝围			袖肥		
裤口			袖口		
内长			袖长		
拉链长			领围		
外长			领深		
胸围			领宽		
以下几点在生产中必须更正： ＿＿＿＿＿＿ ＿＿＿＿＿＿ ＿＿＿＿＿＿					

生产前需确认：

前身口袋 _____　　　后身口袋 _____

辅料 _____

_____

其他 _____

服装结构

1）　　　　　　尺码　　　　　规格

前身口袋　　　（　　）　　____×____

　　　　　　　（　　）　　____×____

　　　　　　　（　　）　　____×____

2）　　　　　　尺码　　　　　规格

后身口袋　　　（　　）　　____×____

　　　　　　　（　　）　　____×____

　　　　　　　（　　）　　____×____

3）后育克尺寸 _____（于后裆缝口处）_____（于侧缝处）

4）后身口袋位置

（a）后育克下 _____，且与后育克底边平行

（b）后腰头下 _____，且与腰头底边平行

（c）腰头底边下 _____（于后裆缝处）

_____（于侧缝处）

颜色搭配

裤/衣身　　　明线　　　装饰物　　　商标　　　金属制品

_____　　_____　　_____　　_____　　_____

面料

名称/编号 _____　面料成分 _____

重量 _____　　　盎司/平方英尺

_____　　　磅/打

结构 _____×_____

英支×　　　　英支

辅料

	可接受	不可接受
面线粗细	（　　）	（　　）
纽扣/按扣	（　　）	（　　）
铆钉	（　　）	（　　）
拉链	（　　）	（　　）
衬料	（　　）	（　　）
里料	（　　）	（　　）
绳边/内置物	（　　）	（　　）
罗纹	（　　）	（　　）
束带（抽绳）	（　　）	（　　）
肩垫	（　　）	（　　）

Garment Foreign Trade Documentary Translation
服装外贸单据翻译

Chapter 8　Manufacturing Contract and Purchase Order
第八章　加工合同和采购订单

项目名称：加工合同和采购订单

项目内容：1．加工合同

2．采购订单

教学学时：6课时

教学目的：让学生了解外贸加工合同和采购订单的一般格式和内容；培养学生对外贸加工合同和采购订单的阅读能力和翻译能力。

教学方式：由教师通过课前导入任务引出本章内容。分析讲解加工订单和采购订单的格式、注意要点和翻译。通过拓展外贸案例，深入学习加工订单和采购订单的翻译。总结归纳学习难点和重点，组织学生进行实操练习。

教学要求：（1）掌握本课词汇。

（2）理解加工合同、采购订单的内容和签订要求。

（3）熟练阅读、翻译英文加工合同和采购订单。

Chapter 8
Manufacturing Contract and Purchase Order

Part 1   Manufacturing Contract

The manufacturing contract (Table 8-1) should contain the following information:

(1) Serial number.

(2) Ordered date.

(3) Delivery date or total ordered amounts.

(4) Where and how to be delivered.

(5) The prices and terms of order.

(6) Size and color break down.

(7) Packing and shipping instructions.

Part 2   Purchase Order

The purchase order (Table 8-2) is made out by purchasing department. It should contain the following information:

(1) Ordered date.

(2) Contract number.

(3) Production order number.

(4) Delivery date.

(5) Company division.

(6) Full descriptions on color, style, quantity and specifications.

(7) Prices and terms.

(8) Where and how to be shipped together with packing instructions.

Five copies of purchase order are distributed to purchasing department, Vendor, Account

Department, Shipping Department and Sales Department. Advantages of this form are easy to follow up and complain or claim according to the purchase orders specification. If the details are not clearly shown on purchase order, company will sometimes suffer loss.

Table 8-1　Manufacturing Contract

HONGDA Trading Ltd.	
HUIDA GARMENT CO.,LTD. Building, 2/F, NO.8, ZHAODONG ROAD, FOSHAN, CANTON PROVINCE, CHINA.	16/F Nan Hua No. 25-28 Nathan Road, Kowloon, Hong Kong

<hereinafter called the Seller/Manufacture>

MANUFACTURING CONTRACT NO.: MC18-09003W　DATE: 20-Aug.-2018

We, HongDa Trading Ltd., Hong Kong hereby agrees to buy and the Seller/Manufacture agrees to sell and /or manufacture the following merchandise(s) on the terms and conditions that stipulated in the General Conditions of Purchase and /or Sub-contraction:

STYLE, SIZE AND ASSORTMENT BREAKDOWN AS PER PRODUCTION INFORMATION SHEET WHICH ARE INTEGRAL PARTS OF THIS CONTRACT.

DESCRIPTION: MEN'S L/S SHIRT　CATEGORY NO.: USA CAT.640

FABRIC: T/R BENGALINE WOVEN FABRIC

COLOR/PATTERN: WHITE #8JS0165

SIZE BREAKDOWN:

BUYER	PO#	STYLE#	SIZE RATIO	S	M	L	XL	TTL
				2	4	4	2	12
TARGET	8593TA	8HT4085WHY2		190	380	380	190	1,140 PCS.

UNIT PRICE: 　HK$　　　/PC.	TOTAL QUANTITY: 　　　1,140 PCS.	TOTAL AMOUNT: 　　　HK$

TERMS: FOB

LABEL: "EXCETERA"

SHIPMENT: ON/BEFORE 1-OCT.-2018 EX-CHINA BY SEA TO U.S.A. PORT.

PAYMENT: BY T.T. PAYMENT ONE WEEK AFTER RECEIVED FULL SET OF NEGOTIABLE SHIPPING DOCUMENT.

REMARKS:

(1) TOLERANCE SHIPMENT OVER +/-5% PER COLOR & PER SIZE IS NOT ALLOWED.

(2) HANGER PACK, EACH PIECE ON HANGER IN A POLYBAG, SIX PCS. TO AN INNER CASE IN SOLOD STYLE ASSORTED SIZES, TWELVE PCS. TO A MASTER CARTON IN SOLID STYLE ASSORTED SIZES.

　＊＊＊NO PINS, NO CLIPS, NO INDIVIDUAL POLYBAGS, NO TISSUE PAPER＊＊＊

(3) PAYMENT WILL BE DELAYED IF THIS CONTRACT IS NOT SIGNED & RETURNED TO US BEFORE SHIPMENT.

(4) SAMPLES:

PRE-PRODUCTION SAMPLE: 2 PCS. IN SIZE M(MUST BE SUBMITTED BEFORE PRODUCTION)

Continued

SALESMAN SAMPLE: 3 PCS. IN SIZE L(MUST BE SUBMITTED BEGORE 15-Sept.-2018)

SHIPMENT SAMPLE: 2 PCS. IN SIZE L(MUST BE SUBMITTED BEGORE 15-Sept.-2018)

　　* ALL SAMPLES SHOULD BE SUBMITTED BY HONGDA TRADING LTD. AT FREE OF CHARGE BASIS. *

For and on behalf of

Hong Da Trading Ltd.   Seller/Manufacturer:

_____   _____

Authorized Signature(s)   Authorized Signature(s)

Table 8-2　Purchase Order

HONGWEI CORP. PURCHASE ORDER

Page: 1

P.O.NO.:	CONTRACT NO.:	COMPANY DIVISION: HONGWEI	STYLE NO.: S-47803	DATE: 4/26/2018

MANUFACTURER:
MEI MEI CORP. (USA)

ADDRESS:
200 PARK AVENUE (PAN AM BUILDING)
NEW YORK N.Y. 10166-0130
U.S.A

CHECK GARMENT STRUCTURE
☐ KNIT
☐ WOVEN
☑ KNIT/WOVEN

CHECK FABRIC
☐ 65% POLYESTER
☐ 35% COTTON
☐ 100% COTTON
☐ 100% ACRYLIC
☑ OTHER:

TOPS:　55% COTTON
　　　　45% POLYESTER
SHORTS:　50% POLYESTER
　　　　　50% COTTON

	VESTS	TOPS	PANTS
GARMENT TYPE			
NET WEIGHT (AVERAGE SIZE) (MIN.)	SIZE___	SIZE___	SIZE___
	___LB./DZ.	___LB./DZ.	___LB./DZ.
TOTAL CARTON	TOTAL QUANTITY		F.O.B PER DZ.
4,000 CTNS.	8,000 DZ.		US$ 28.40 (L.D.P)
TOTAL COST OF ORDER	TERM		PORT
US$ 227,200.00	F.O.B		HAITI
DELIVERY DATE			CANCEL DATE
JUL.-SEP. 2018			—

SHIP VIA　DESTINATION
☑ SEA　NEW YORK
☐ AIR

DETAILED SPECIFICATIONS OF GARMENT

BOY'S 2-4 KNIT/WOVEN SHORT SET.

SIZE	2T	3T	4T
RATIO	3	5	4

☐ SOLID COLOR

☑ ASSORTED COLORS

☑ ASSORTED SIZES

(MAX.)CARTON MEASUREMENT	
SIZE	CU.FT.

Continued

Page: 2

PACKING	HANGER PACK	ONE PC. PER ONE POLYBAG, ONE DZ. PER ONE INNER BOX, PACK A+B=2DZ. PER ONE EXPOR CARTON.

STYLE NO.: S-47803 QTY.: 8,000 DZ.   STYLE NO.: _____ QTY.: _____ DZ.

	SIZE	2T	3T	4T			SIZE	
PACK A					PACK A			
STY. A		1	3	2				
STY. B		2	2	2				
	TOTAL: 12 PCS.					TOTAL:		

	SIZE	2T	3T	4T			SIZE	
PACK B					PACK B			
STY A		1	3	2				
STY B		2	2	2				
	TOTAL: 12 PCS.					TOTAL:		

POLYBAG	THICKNESS 0.35mm	COUNTRY OF ORIGIN HAITI ☐	LOGO ☐ PLAIN	SIZE ☐ 2T-4T	FIBRE CONTENT ☐	RN#❶ ☐ 46795	WASHING/CARE INSTRUCTION+WARNING ☐	ADVISE LATER ☐

HANGER	COLOR WHITE ☐ BLACK ☐ OTHER ☐	TOPS/VESTS HANGER ☐ PANTS HANGER ☐	SIZE 2-4 K-10" HANGER / SIZE __ HANGER / SIZE __ HANGER / SIZE __ HANGER	SIZE SPECIFICATION ☐ ALREADY PROVIDED ☐ ADVISE LATER ☐

MAIN LABEL LITTLE BY LITTE SEWIN AT BACK NECK & BACK RISE

CARE LABEL ☐				
SIZE LABE ☐ ADVISE LATTER ☐	FIBRE CONTENT ☐	WOVEN ☐ PRINT ☐	COUNTRY OF ORIGN ☐	WASHING/CARE INSTRUCTION ☐

PRICE TICKET ☐	LOCATION SWIFT TACK ☐ STRING ☐	ADVICE LATER ☐ /SET/PCS. ☐	HANG TAG ☐ LBYL ☐	LOCATION NECK LABEL SWIFT TACK ☐ STRING ☐	ADVICE LATER ☐ /SET/PCS. ☐

❶ RN#represents Registered identification Number.

Continued

Page: 3

APPROVAL SAMPLE	PRE-PRODUCTION SAMPLE	SHIPMENT SAMPLE
2 SET/PCS. SIZE 3T AVAILABLE FABRIC AND COLOR	2 SET/PCS./SIZE 3T PRIOR BULK PRODUCTION	□ PLEASE SUBMIT FULL SIZE/COLOR RANGE WHEN SHIPMENT EFFECT

EXPORT CARTON MARKING

EXPORT CARTON □ 3 PLY

INNER CARTON MARKING
INNER CARTON □ 2 PLY

FRONT & BACK

L O G O

◇ S-47803

NEW YORK
MADE IN HAITI
C/NO.: 1-UP

SIDE

LOCO: N.Y.
STY.: NO.S-47803
SIZE: 2T-4T
CONTENT: 2 DZ.
NT.WT.:
GR.WT.:
MEAS.:

ADVISE LATER □

ADVISE LATER □

REMARKS:

ALL FABRIC AND ACCESSORIES MUST BE APPROVED BY HONG WEI OFFICE BEFORE BULK PRODUCTION.

New Words and Expressions

1. break down　分析；分类
2. instruction　指令；命令
3. purchase　购买
4. vendor; vender　卖方；卖主
5. advantage　益处；优点
6. follow up　追逐；追求
7. complaint　抱怨；不满；控诉
8. claim　根据保险合约所要求的赔款；要求权
9. hereby　借此；由此
10. merchandise　商品
11. therein　在那方面；在那一点上
12. stipulate　规定；约定
13. sub-contraction　托外（协）加工
14. assortment　搭配；配套商品
15. integral　构成整体所需要的
16. category　（整个系统或组合中的）部门；范畴
17. bengaline　罗缎
18. pattern　花样；式样
19. F.O.B (free on board)　离岸交货价；船上交货价
20. negotiable　可商议的；可谈判处理的
21. document　文件；公文；证件
22. polybag　胶袋
23. solid　纯的；全部一样的
24. assorted　各种各样的；混合的
25. pin　大头针；扣针；包装针
26. clip　胶夹；别针；首饰别针
27. tissue paper　拷贝纸；薄纸
28. submit　建议；主张；服从；批准
29. salesman sample　展销板
30. shipment sample　装船样板
31. free　免费的
32. on behalf of　为了……的利益；代表
33. authorize　批准；许可
34. pant　裤子；长裤
35. average　平均数；平均的
36. CTNS (cartons)　箱数
37. polyester　涤纶
38. acrylic　腈纶
39. cancel　删去；注销；取消
40. ratio　比；比率
41. plain　平纹的；素色的；平纹布
42. logo　标识
43. hangtag　吊牌
44. tack　粗缝；假缝；松散或临时地系、扎、扣、栓
45. range　范围；色阶；色程

Exercises

1. List the content of manufacturing contract.

2. List the content of the purchase order.

3. Translate the contract below.

GUANGZHOU TEXTLES
IMPORT & EXPORT CORPORATION

(ORIGINAL)

89 Wen Deng Road, Guangzhou, China

Contract No.

The Buyers:

CONTRACT

Date:

FAX: 020–291730

Telex number: TEXTILE

This CONTRACT is made by and between the Buyers and the Sellers; whereby the Buyers agree to buy and the Sellers agree to sell the undermentioned goods on the terms and conditions stated below:

(1)	(2)	(3)	(4)	(5)
Name of Commodity, Specifications, Packing Terms and Shipping Marks	Quantity	Unit Price	Total Amount	Time of Shipment

(6) Port of Loading.

(7) Port of Destination.

(8) Terms of Payment:Upon receipt from the Sellers of the advice as to the time and quantify expected ready for shipment, the Buyers shall open, 20 days before shipment, with the Bank of China, Shanghai, an irrevocable Letter of Credit in favour of the Sellers payable by the opening bank against sight draft accompanied by the documents as stipulated in Clause (9) of this Contract.

(9) Documents:To facilitate the Buyers to check up, all documents should be made in a version identical to that used in this contract.

(10) Terms of Shipment（FOB Delivery）:

For the goods ordered in this Contract, the carrying vessel shall be arranged by the Buyers or the Buyers' Shipping Agent China National Chartering Corporation. The Sellers shall bear all the charges and risks until the goods are effectively loaded on board the carrying vessel.

The Buyers:

The Sellers:

第八章　加工合同和采购订单

第一节　加工合同

加工合同（表8-1）必须包括下列内容：

（1）序号。

（2）订单签订日期。

（3）交货日期和订购总额。

（4）交货地点和方式。

（5）价格和条件。

（6）尺寸和颜色搭配。

（7）包装和装运要求。

第二节　采购订单

采购订单（表8-2）由采购部门开出。采购订单应包括下列内容：

（1）订单签订日期。

（2）合同编号。

（3）制造通知单号。

（4）交货日期。

（5）公司负责部门。

（6）有关颜色、款式、数量和规格尺寸的要求。

（7）价格和条件。

（8）装运地点和方式等。

　　五份采购订单复印件将会分派给采购部门、卖方、财务部门、运输部门和销售部门。这种表格的好处是易于进行进度检查、投诉或根据订单索赔。如果采购订单上的资料不清楚，会导致公司遭受损失。

表8-1　加工合同

晖达制衣有限公司
中国广东省佛山市
朝东路8号2层

宏达贸易有限公司
香港九龙弥敦道25-28号
南华大厦16层

<以下统称卖方/制造商>

制造合同号：MC98-09003W	日期：20-08-2018

我方，中国香港宏达贸易有限公司同意买，卖方／制造商同意出售和（或）加工下列商品，并执行约定的采购和（或）委托外加工的一般条件。

生产信息单中显示的款式、尺码和搭配等都是此合同的构成部分。

款式：男装长袖恤衫　　　　　　　　种类编号：USA CAT.640

布料：涤纶/黏胶　罗缎机织布

颜色/花样：白色 #8JS0165

尺码搭配：

买方	订单号	款号	尺码	S	M	L	XL	共计
			比率	2	4	4	2	12
目标	8593TA	8HT4085WHY2		190	380	380	190	1,140件

单价：HK$　　/件	总数量：1,140件	总金额：港币（HK$）

报价：离岸交货价

商标：“EXCETERA”

交货期：在2018年10月1日或之前海运到美国港口。

付款方式：收到整套的装运文件后一星期内按T.T.方式付款。

备注：（1）每色和每码短装或溢装超过5%的不能接受。

（2）带衣架包装，各件挂于衣架上并装入胶袋。

6件单款分码装入1内箱。

12件单款分码装入1外箱。

＊＊＊不用包装针、胶夹、另外的胶袋及拷贝纸＊＊＊

（3）如果装运之前此合同未署名并寄回我方，将延迟付款。

（4）样板（样衣）：

产前样：M码2件（大货生产前必须经确认）

展销样：L码3件（必须在2018年9月15日前确认）

（装）船样：L码2件（必须在2018年9月15日前确认）

＊所有样板（样衣）须经宏达贸易有限公司确认。样板（样衣）免费。

宏达贸易有限公司：

＿＿＿＿＿＿＿＿
（代表签字）

卖方/制造商：

＿＿＿＿＿＿＿＿
（代表签字）

表8-2 采购订单

宏威有限公司采购单

第1页

订单号：	合同号：	公司负责部门：宏威	款号：S-47803　　订单签订日期：4/26/2018

制造商： MEI MEI 公司（美国）	服装类型 服装构成 □ 针织 □ 机织 □ 针织/机织	背心　尺码　———　磅/打	上装　尺码　———　磅/打	裤子　尺码　———　磅/打
		重 （均码） （最小量）		

	面料	总箱数 4,000箱	总数量 8,000打	离岸交货价/打 US$ 28.40
	□ 65% 涤纶 □ 35% 棉 □ 100% 棉 □ 100% 腈纶 ☑ 其他： 　上装：　55% 棉　45% 涤纶 　短裤：　50% 涤纶　50% 棉	订单总价值 US$ 227,200.00	运输条款	港口 HAITI
			离岸交货价	

地址： 200 PARK AVENUE (PAN AM BUILDING) NEW YORK N.Y. 10166-0130 U.S.A	目的地 纽约	交货日期 2018年7～9月	取消日期 —

运输： ☑ 海运 □ 空运	服装细节说明 男童2～4码针织/机织短套装	□ 单色　☑ 混色　☑ 混码	（最大）外箱尺寸 尺码 立方英尺

尺码	2T	3T	4T
比率	3	5	4

续表

第2页

包装要求	带衣架包装						一件入一胶袋，一打入一内盒，包装A+B=2打入一外箱		
	款号：S-47803　数量：8,000 打					款号：　　　　数量：　　　　打			
包装方法 A	尺码	2T	3T	4T		包装方法 A	尺码		
款式 A		1	3	2					
款式 B		2	2	2					
		共 12 件					共		
包装方法 B	尺码	2T	3T	4T		包装方法 B	尺码		
款式 A		1	3	2					
款式 B		2	2	2					
		共 12 件					共		

胶袋	厚度 0.35 mm	标识 平纹	产地 HAITI	尺码 2T-4T	纤维成分 □	公司注册号 46795 □	洗水养护 说明+警告		稍后说明 □
	颜色 白色 □ 黑色 □ 其他 □								

| 衣架 | | 上装/背心衣架 □ 裤子衣架 □ | | 稍后说明 □ | | 尺码 2-4　K-10"　衣架 尺码 　—　　衣架 尺码 　—　　衣架 尺码 　—　　衣架 | 规格尺寸表 已提供 □ 稍后告知 □ | | |

| 主唛（商标） 水洗标 □ 尺码标 □ | LITTLE BY LITTE | 车缝于 后领和后档 □ | | 位置 □ 胶针 □ 绳 | 稍后说明 □ | 纤维成分 机织 □ 打印 □ | | 洗水养护说明 位置：后领商标处 □ 胶针 □ 绳 | 产地 □ |

| 价格牌 □ | | | | 稍后说明 □ 套/件 □ | | 吊牌 □ LBYL | | | 稍后说明 □ 套/件 □ |

续表

	第3页

外箱标注（唛头）　　　　　　　　外箱 □ 3层

正面和背面　　　　　　　　侧面

```
  L    O    G    O

      ◇ S-47803 ◇

NEW        YORK
  MADE IN HAITI
  C/NO: 1-UP
```

LOCO: N.Y.
款号: S-47803
尺码: 2T～4T
数量: 2打
净重:
毛重:
尺寸:

内箱标注
内箱 □ 2层

稍后说明 □

稍后说明 □

备注:

在大货生产之前，所有面料和辅料须经宏威确认。

本模块微课资源（扫描二维码观看视频）

5.1 English Expression of Tops　　5.2 English Expression of Bottoms　　5.3 English Expression of Parts of Pants　　5.4 Dresses　　5.5 Shirts

5.6 Collar　　5.7 Pocket　　5.8 Yoke　　5.9 "褶" in Garment　　6.1 Sample in Garment

7.1 English Expression of Measurement Unit in Garment　　7.2 English Expression of Stitches　　8.1 OEM & ODM & OBM　　8.2 Labels in Garment

Module 3
Garment Producing and Packaging

模块三
服装生产和包装

Garment Producing and Packaging
服装生产和包装

Chapter 9　Abbreviation for Garment
第九章　服装英语的缩写

项目名称： 服装英语的缩写

项目内容： 1. 缩写的目的

2. 服装外贸缩写

教学学时： 4课时

教学目的： 让学生认识和掌握服装外贸中专业术语的缩写，了解其在服装外贸中的作用，培养学生掌握基本的服装外贸知识，并能熟练运用。

教学方式： 通过课前任务，引入本章内容；分析讲解服装英语术语缩写目的和分类；通过拓展外贸案例，让学生深入了解服装英语术语缩写；最后对本章学习难点和重点进行归纳，学生通过实操练习进一步熟悉和掌握本章节内容。

教学要求： （1）认识服装英语专业术语的缩写。

（2）了解服装专业术语缩写在服装外贸中的作用。

（3）通过实操练习熟悉和掌握服装英语缩写。

Chapter 9   Abbreviation for Garment

Part 1   Purposes of Abbreviations

As foreign trade practitioners, we often find that foreign customers are used to abbreviations in our communication with customers. The use of abbreviations in foreign trade has the following functions:

(1) Brief and easy.

(2) Effective and time-saving.

(3) Habits of foreign customers.

(4) Conventional expressions between foreign trade experts.

Part 2   Abbreviations in Garment Foreign Trade

Abbreviations frequently used by foreign customers in garment foreign trade mainly include:Abbreviation of Garment Daily English, Abbreviation of Garment Parts, Abbreviation of Fabrics, Abbreviation of Colors, Abbreviation in Foreign Trade Documents , Abbreviation of Quantity Unit and Abbreviation of Size (Table 9-1 to Table 9-7).

Table 9-1   Abbreviations of Garment Foreign Trade Daily English

No.	Abbr.	Full Name	No.	Abbr.	Full Name
1	BTW	by the way	7	BBL	be back later
2	IMO	in my opinion	8	ATM	at the moment
3	FYI	for your information	9	CYL	see you later
4	IKR	i know right	10	J/K	just kidding
5	AFAIK	as far as i know	11	NP	no problen
6	IMHO	in my humble opinion	12	ASAP	as soon as possible

Table 9-2 Abbreviations of Garment Parts

No.	Abbr.	Full Name	No.	Abbr.	Full Name
1	A.H.	armhole	41	K.	knit
2	B.	bust	42	K.L.	knee line
3	B.	back	43	N.-W.	nape to waist
4	B.C.	biceps circumference	44	N.H.	neck hole
5	B.H.	button hole	45	NK	neck
6	B.L.	back length	46	N.L.	neck length
7	B.L.	bow line	47	N.L.	neck line
8	B.N.	back neck	48	N.P.	neck point
9	B.P.	bust point	49	N.R.	neck rib
10	BR.	back rise	50	N.S.	neck size
11	B.S.L.	back shoulder line	51	N.S.P.	neck shoulder point
12	BSP.	back shoulder point	52	N.W.L.	neck waist length
13	BTM.	bottom	53	O.S.	outside seam
14	BTN.	button	54	OVRLK.	overlock
15	B.W.	back width	55	PKT.	pocket
16	C.	chest	56	S.	sleeve
17	C/B (C.B.)	center back	57	S.A.	seam allowance
18	CBF.	center back fold	58	S.B.	single breasted
19	CBL.	center back line	59	S.C.	stand collar
20	C/F (C.F.)	center front	60	S.D.	scye depth
21	CFL.	center front fold	61	SGL.NDL.	single needle
22	C.P.L.	collar point length	62	S.L.	sleeve length
23	C.P.W.	collar point width	63	SNL.	single
24	C.W.	cuff width	64	S.P.	shoulder point
25	D.B.	double breasted	65	S.P.I.	stitch per inch
26	E.C.	elbow circumference	66	S.P.M.	stitch per minutes
27	E.L.	elbow line	67	S.S.	sleeve slope
28	EMB.	embroidery	68	S.S.P.	houlder sleeve point
29	E.P.	elbow point	69	S.T.	sleeve top
30	FAB.	fabric	70	S.W.	shoulder width
31	F.L.	front length	71	T/S	top stitches
32	F.N.	front neck	72	T.L.	trousers length
33	F.N.P.	front neck point	73	T.R.	trouser rise
34	F.R.	front rise	74	T.S.	thigh size
35	F.S.	fist size	75	UBL.	under bust line
36	F.W.	front width	76	W.	woven
37	H.S.	head size	77	W.	waist
38	I.	inseam	78	W.	width
39	LBL.	label	79	W.B.	waistband
40	LOA.	length over all	80	W.L.	waist line

Table 9-3 Abbreviations of Fabrics

No.	Abbr.	Full Name	No.	Abbr.	Full Name
1	A.	acrylic	13	PV.	polyvinyl
2	AL.	alpace	14	R.	rayon
3	C.	cotton	15	RH.	rabbit hair
4	CH.	camel hair	16	SP.	spandex
5	LY.	lycra	17	S.	silk
6	LA.	lambswool	18	TEL.	tencel
7	L.	linen	19	T.	polyester
8	M.	mohair	20	V.	viscose
9	MD.	modal	21	W.	wool
10	N.	nylon	22	WS.	cashmere
11	PA.	polyamide	23	WA.	angora
12	PP.	polypropylene	24	YH.	yark hair

Table 9-4 Abbreviations of Colors

No.	Abbr.	Full Name	No.	Abbr.	Full Name
1	BU	blue	18	RO.	reddish orange
2	BK	black	19	RY.	reddish yellow
3	BN.	brown	20	RP.	red purple
4	GN	green	21	YO.	yellowish orange
5	GY.	grey	22	YR.	yellowish red
6	OG.	orange	23	YG.	yellow green
7	PK.	pink	24	YG.	yellowish green
8	RD.	red	25	FLUO.	fluorescence
9	VT.	violet	26	DK.	dark
10	WH	white	27	LT.	light
11	YE.	yellow	28	CR.	clear
12	BG.	blue green	29	DP.	deep
13	BG.	blueish green	30	P.	pale
14	GY.	greenish yellow	31	B.	bright
15	GB.	greenish blue	32	D.	dull
16	PB.	purplish blue	33	S.	strong
17	PR.	purplish red	34	V.	vivid

Table 9-5   Abbreviations in Foreign Trade Documents

No.	Abbr.	Full Name	No.	Abbr.	Full Name
1	T.T.T	tone to tone	27	FB.	freight bill
2	DTM	dye to match	28	CAT.	catalogue
3	AOP	all over print	29	DS	detail sketch
4	DTMSA	dye to match surrounding area	30	C/NO., CTN. NO.	carton no.
5	SPEC.	specification	31	G.W.	gross weight
6	LCL	less container loaded	32	D/Y	delivery
7	CY	container yard	33	L/D	lab dip
8	QC	quality control	34	CI	corporate identify
9	W/R	waterproof	35	AWB. NO.	air way bill no.
10	Y/D	yarn dye	36	BOM	bill of material
11	W/	with	37	BX.	boxes
12	M/C	machine	38	MAT.	material
13	MAT.	material	39	MEAS.	measurement
14	C.F., C & F	coat and freight	40	PKG.	package
15	CIF	cost, insurance & freight	41	FOB	free on board
16	C/D.	certificate of delivered	42	C.&D.	collected and delivered
17	MKT.	market	43	C.A.D.	cash against documents
18	W.I.P.	work in process	44	C/O, C.O.	country of origin
19	WMSP.	workmanship	45	R.T.W.	ready to wear
20	TOL	tolerance	46	L/H	lable & hangtag
21	A/C	account	47	P.P.	paper pattern
22	A/W, AW	actual weight	48	PB.	private brand
23	ADD.	address	49	PC.	price
24	AGT.	agent	50	QTY.	quantity
25	AMT.	amount	51	QLY.	quality
26	CS.	commercial standards	52	RN.	reference number

Table 9-6   Abbreviations of Quantity Unit

No.	Abbr.	Full Name	No.	Abbr.	Full Name
1	K.	kilo	8	GAL.	gallon
2	PC	piece	9	IN.	inch
3	PCS.	pieces	10	FT.	feet
4	L.	ligne	11	SQ.	square
5	LB.	libra（e）	12	CBM	cubic meter
6	OZ.	ounce	13	YD.	yard
7	DOZ.	dozen	14	DBL.	double

Table 9-7   Abbreviations of Size

No.	Abbr.	Full Name	No.	Abbr.	Full Name
1	S	small	5	P.	petite
2	M	medium	6	NB	new birth
3	L	large	7	M	month
4	XL	extra large	8	Y	year

New Words and Expressions

1. fluorescence   荧光
2. overlock   包缝
3. water proof   防水
4. yarn dye   色织
5. work in process   半成品
6. ready to wear   成衣
7. workmanship   手工；车工
8. tolerance   允许
9. account   账单；账目
10. agent   代理商；代理人
11. catalogue   产品目录
12. detail sketch   细节图
13. freight bill   装货清单
14. delivery   出货，交付
15. lab dip   色卡；色样
16. paper pattern   纸样
17. private brand   个人商标
18. reference number   参考号
19. ligne   莱尼/号（纽扣大小单位）
20. libra（e）   磅
21. ounce   盎司、安士
22. dozen   打
23. gallon   加仑
24. inch   英寸
25. feet   英尺
26. square   平方
27. cubic meter   立方米
28. yard   码

Exercises

Translate the following emails ,especially pat attention to the abbreviation in it.

Email 1

> MR. DANNY,
>
> WELL RCVD THE CK TODAY, THX VERY MUCH FOR THE UNDERSTANDING AND HELP.
>
> MEANWHILE, PLS KINDLY REFER TO BELOW MSG FM MY QINGDAO OFFICE ABT HOLDING THE 2ND 5 CNTRS & 3RD 2 CNTRS ALL TO MIDLAND, TX DUE TO THE PAYMENT BETWEEN YR SIDE & THE FACTORY. SEVERAL CNTRS WILL ARRIVE IN TX DURING THIS WEEK N THESE ARE URGENT SHPT, SO PLS URGENTLY.
>
> HOPE TO TALK TO THE FACTORY SO THAT WE CAN RLS THE SHPT W/O ANY DELAY.
>
> THX!
>
> MARIE

Email 2

> MARIE,
>
> LEARNT FROM S/ THAT PART OF CARGO VALUE, RMB600,000 WAS NOT PAID BY CNEE, SO UNTIL NOW S/ DO NOT PAY US FOR THE O/F AND LOCAL CHARGE. IT'S A VERY BIG AMOUNT.
>
> KINDLY PLS ADVISE IF THE CARGO UNDER MB/L NO.:OCLTAO30000006A ARRIVE THE DEST.IF NOT, COULD YOU PLEASE HOLD THE CARGO UNTIL OUR FURTHER INFM?
>
> ALSO PLS HOLD THE CARGO OF MB/L NO.: POCLTAO30000007A!
>
> MANY TKS!
>
> B.RGDS,
>
> LESLEY LAI

References:

S: sale, O/F: ocean freight, CNEE: consignee, MB/L: master bill of lading, INFM: information.

第九章　服装英语的缩写

第一节　缩写的目的

作为外贸从业人员，我们在与客户沟通中，经常会发现国外客户习惯使用缩写。在外贸中使用缩写具有以下作用：

（1）简洁方便。

（2）高效省时。

（3）国外客户的习惯。

（4）外贸熟手之间约定俗成的表达。

第二节　服装外贸缩写

在服装外贸中，国外客户经常使用的缩写，主要包含：服装日常用语缩写、服装部位缩写、面料纤维缩写、颜色缩写、外贸单证中的缩写、数量单位的缩写以及尺码的缩写（表9-1～表9-7）。

表9-1　服装外贸日常用语缩写

序号	缩写	全称	中文	序号	缩写	全称	中文
1	BTW	by the way	顺便提下	7	BBL	be back later	待会回来
2	IMO	in my opinion	依我所见	8	ATM	at the moment	目前
3	FYI	for your information	供参考	9	CYL	see you later	回见
4	IKR	i know right	本人知悉	10	J/K	just kidding	开玩笑
5	AFAIK	as far as i know	以我所知	11	NP	no problem	没问题
6	IMHO	in my humble opinion	依我拙见	12	ASAP	as soon as possible	尽快

表9-2　服装部位缩写

序号	缩写	全称	中文	序号	缩写	全称	中文
1	A.H.	armhole	袖窿	29	E.P.	elbow point	肘点
2	B.	bust	胸围	30	FAB.	fabric	布料
3	B.	back	后	31	F.L.	front length	前长
4	B.C.	biceps circumference	上臂围	32	F.N.	front neck	前领围
5	B.H.	button hole	纽门、扣眼	33	F.N.P.	front neck point	前颈点
6	B.L.	back length	后长	34	F.R.	front rise	前裆
7	B.L.	bow line	肘线	35	F.S.	fist size	手头围
8	B.N.	back neck	后领围	36	F.W.	front width	前胸宽
9	B.P.	bust point	胸（高）点	37	H.S.	head size	头围
10	BR.	back rise	后裆	38	I.	inseam	内长
11	B.S.L.	back shoulder line	后肩线	39	LBL.	label	唛头、商标
12	BSP.	back shoulder point	后肩颈点	40	LOA.	length over all	全长
13	BTM.	bottom	衫脚	41	K.	knit	针织
14	BTN.	button	纽扣	42	K.L.	knee line	膝围线
15	B.W.	back width	后背宽	43	N.-W.	nape to waist	腰直
16	C.	chest	胸围	44	N.H.	neck hole	领圈、领口
17	C/B (C.B.)	center back	后中	45	NK	neck	颈圈
18	CBF.	center back fold	后中对折	46	N.L.	neck length	领长
19	CBL.	center back line	后中线	47	N.L.	neck line	领围/线
20	C/F (C.F.)	center front	前中	48	N.P.	neck point	颈点、肩顶
21	CFL.	center front fold	前中对折	49	N.R.	neck rib	领高
22	C.P.L.	collar point length	领尖长	50	N.S.	neck size	颈围
23	C.P.W.	collar point width	领尖宽	51	N.S.P.	neck shoulder point	颈肩点
24	C.W.	cuff width	袖口宽	52	N.W.L.	neck waist length	背长
25	D.B.	double breasted	双排纽扣，双襟	53	O.S.	outside seam	外缝长
26	E.C.	elbow circumference	肘围	54	OVRLK.	overlock	及骨、包缝
27	E.L.	elbow line	手肘线	55	PKT.	pocket	口袋
28	EMB.	embroidery	刺绣	56	S.	sleeve	袖长、袖子

续表

序号	缩写	全称	中文	序号	缩写	全称	中文
57	S.A.	seam allowance	止口	69	S.T.	sleeve top	袖山
58	S.B.	single breasted	单排纽扣、单襟	70	S.W.	shoulder width	肩宽
59	S.C.	stand collar	领座	71	T/S	top stitches	面线
60	S.D.	scye depth	腋深	72	T.L.	trousers length	裤长
61	SGL.NDL.	single needle	单针	73	T.R.	trouser rise	裤直/裆
62	S.L.	sleeve length	袖长	74	T.S.	thigh size	腿围
63	SNL.	single	单线	75	UBL.	under bust line	下胸围线
64	S.P.	shoulder point	肩高点	76	W.	woven	机织
65	S.P.I.	stitch per inch	每英寸线迹数	77	W.	waist	腰围
66	S.P.M.	stitch per minutes	每分钟线迹数	78	W.	width	宽度
67	S.S.	sleeve slope	肩斜	79	W.B.	waistband	腰头
68	S.S.P.	houlder sleeve point	肩袖点	80	W.L.	waist line	腰线

表9-3　面料纤维缩写

序号	缩写	全称	中文	序号	缩写	全称	中文
1	A.	acrylic	腈纶	13	PV.	polyvinyl	维纶
2	AL.	alpace	羊驼毛	14	R.	rayon	人造棉
3	C.	cotton	棉	15	RH.	rabbit hair	兔毛
4	CH.	camel hair	驼毛/绒	16	SP.	spandex	氨纶
5	LY.	lycra	莱卡	17	S.	silk	丝
6	LA.	lambswool	羊羔毛	18	TEL.	tencel	天丝
7	L.	linen	亚麻布	19	T.	polyester	涤纶
8	M.	mohair	马海毛	20	V.	viscose	黏胶
9	MD.	modal	莫代尔	21	W.	wool	羊毛
10	N.	nylon	锦纶、尼龙	22	WS.	cashmere	羊绒
11	PA	polyamide	聚酰胺（尼龙）	23	WA.	angora	山羊毛
12	PP.	polypropylene	丙纶	24	YH.	yark hair	牦牛毛

表9-4 颜色的缩写

序号	缩写	全称	中文	序号	缩写	全称	中文
1	BU	blue	蓝色	17	RO.	reddish orange	红橙色
2	BK	black	黑色	18	RY.	reddish yellow	红黄色
3	BN.	brown	棕色	19	RP.	reddish purple	红紫色
4	GN	green	绿色	20	YO.	yellowish orange	黄橙色
5	GY.	gray	灰色	21	YG.	yellow green	黄绿色
6	OG.	orange	橙色	22	YG.	yellowish green	黄绿色
7	PK.	pink	粉红色	23	FLUO.	fluorescence	荧光
8	RD.	red	红色	24	DK.	dark	深色
9	VT.	violet	紫色	25	LT.	light	浅色
10	WH	white	白色	26	CR.	clear	透明
11	YE.	yellow	黄色	27	DP.	deep	深色
12	BG.	blueish green	蓝绿色	28	P.	pale	淡的
13	GY.	greenish yellow	绿黄色	29	B.	bright	明亮
14	GB.	greenish blue	绿蓝色	30	D.	dull	浊的
15	PB.	purplish blue	紫蓝色	31	S.	strong	强烈的
16	PR.	purplish red	紫红色	32	V.	vivid	鲜艳的

表9-5 外贸单据中的缩写

序号	缩写	全称	中文	序号	缩写	全称	中文
1	T.T.T	tone to tone	配色	11	W	with	具有
2	DTM	dye to match	顺色	12	M/C	machine	机械
3	AOP	all over print	印花	13	MAT.	material	物料
4	DTMSA	dye to match surrounding area	配所在部位颜色	14	MKT.	market	市场
5	SPEC.	specification	细则	15	CIF	cost, insurance & freight	到岸价
6	LCL	less container loaded	拼柜	16	C/D.	certificate of delivered	交货证明书
7	CY	container yard	走整柜	17	C.F., C & F	coat and freight	离岸加运费价格
8	QC	quality control	质量控制	18	W.I.P.	work in process	半成品
9	W/R	waterproof	防水	19	WMSP.	workmanship	手工，车工
10	Y/D	yarn dye	色织	20	TOL	tolerance	允许

序号	缩写	全称	中文	序号	缩写	全称	中文
21	A/C	account	账单，账目	37	BX.	boxes	箱，盒
22	A/W	actual weight	实际重量	38	MAT.	material	物料
23	ADD.	address	地址	39	MEAS.	measurement	尺寸
24	AGT.	agent	代理商（人）	40	PKG.	package	包装
25	AMT.	amount	总计，总额	41	FOB	free on board	离岸价
26	CS.	commercial standards	商业标准	42	C.&D.	collected and delivered	货款两清
27	FB.	freight bill	装货清单	43	C.A.D.	cash against documents	凭单据付款
28	CAT.	catalogue	产品目录	44	C/O，C.O.	country of origin	原产国（地）
29	DS	detail sketch	细节图	45	R.T.W.	ready to wear	成衣
30	C（CTN.）/No.	carton no.	箱号	46	L/H	lable & hangtag	唛头和吊牌
31	G.W.	gross weight	毛重	47	p.p.	paper pattern	纸样
32	D/Y	delivery	出货；交付	48	PB.	private brand	个人商标
33	L/D	lab dip	色卡，色样	49	PC.	price	价格
34	CI	corporate identify	企业标识	50	QTY.	quantity	数量
35	AWB. NO.	air way bill no.	运单号	51	QLY.	quality	质量
36	BOM	bill of material	物料表	52	RN.	reference number	参考号

表9-6 服装单位的缩写

序号	缩写	全称	中文	序号	缩写	全称	中文
1	K.	kilo	千	8	GAL.	gallon	加仑
2	PC	piece	只	9	IN.	inch	英寸
3	PCS.	pieces	件，个	10	FT.	feet	英尺
4	L.	ligne	莱尼	11	SQ.	square	平方
5	LB.	libra（e）	磅	12	CBM	cubic meter	立方米
6	OZ.	ounce	盎司、安士	13	YD.	yard	码
7	DOZ.	dozen	打	14	DBL.	double	双

表9-7 服装尺码的缩写

序号	缩写	全称	中文	序号	缩写	全称	中文
1	S	small	小码	5	P.	petite	小码
2	M	medium	中码	6	NB	new birth	新生儿
3	L	large	大码	7	M	month	月
4	XL	extra large	加大码	8	Y	year	岁

Garment Producing and Packaging
服装生产和包装

Chapter 10   Equipment for Garment
第十章   服装设备

项目名称：服装设备

项目内容：1. 测量工具

           2. 描图工具

           3. 熨烫工具

           4. 缝纫工具

           5. 裁剪工具

教学学时：2课时

教学目的：让学生认识和掌握各类服装设备的英文表述，了解其在服装外贸中的作用。

教学方式：通过课前任务，引入本章内容；分析讲解服装设备的基础知识和英文表述；对本章学习难点和重点进行归纳，学生通过实操练习进一步熟悉和掌握本章节内容。

教学要求：（1）认识服装制作中的各种设备。

          （2）掌握各类服装设备的英文表述。

Chapter 10   Equipment for Garment

Commonly used garment equipment tools are: measuring tools, marking tools, pressing tools, sewing tools and cutting tool.

Part 1   Measuring Tools

The commonly used measure tool we called tape measure. The surface of tape measure is smooth and marked with centimeters and inches.Ruler is made of transparent plastic and is marked with centimeters and inches. There are two specifications: 2.5cm (1 inch) wide, 15.2cm (6 inches) long, 5.1cm (2 inches) wide and 45.7cm (18 inches) long. Yardstick: The yardstick has two lengths of 91.4cm (36 inches) and 114.3cm (45 inches), which are made of metal or wood. Skirt marker is used to accurately mark the length of the skirt. It can be used with painting powder or pin.

Part 2   Marking Tools

Tracing wheel transfers the lines on the template to the fabric and avoid yarn hooking. Tracing paper is used to transfer the lines on the pattern to the fabric. Chalk is made of wax or stone. Wax painting powder is used for wool fabrics, and stone painting powder can be used for other fabrics. French curve is a tool for drawing arcs in various parts of garment.

Part 3   Pressing Equipment

Pressing equipment includes steam and dry iron hanging ironing machine, ironing board, sleeve board, point presser and tailor's ham.

Part 4   Sewing Tools

Sewing tools include various sewing machines, such as: lock stitch machine, zigzag machine, overlock machine, safety stitch overlock machine and felling machine. Presser foot is also one of the components in sewing machine. Presser foot is also one of the components in sewing machine, which can be divided into cording presser foot, piping presser foot and gathering presser foot.The cording presser foot is divided into left and right presser feet, which are used to balance both sides and for sewing cord stitch and installing hidden needle seams for zippers. The piping presser foot is used to pipe cloth strip into a tape. The gathering presser foot is used to evenly sew permanent pleats.

In addition, sewing tools also include hand sewing needles, sewing machine needles, pins, thimbles and beeswax.

Part 5   Cutting Tools

Commonly used cutting tools in Garment are bent-handle dressmaker shears, thread clippers, pinking shears and seam ripper.

New Words and Expressions

1. equipment　设备；装备
2. measure　测量
3. mark　标记；记号
4. press　熨烫；按压
5. sew　缝制
6. cut　裁剪
7. centimeter　厘米
8. inch　英寸
9. transparent　透明的
10. yard　码
11. wheel　车轮；轮子
12. zigzag　锯齿形的；Z字形的
13. overlock　锁边；拷边
14. felling machine　锁眼机
15. component　成分；部件
16. oversee　监督；视察
17. cording　嵌线
18. piping　绲边
19. gathering　打褶
20. pin　大头针
21. thimble　顶针
22. beeswax　蜂蜡
23. bent-handle　弯柄
24. shear　剪刀
25. clipper　剪刀；指甲钳
26. pinking shear　花齿剪
27. seam ripper　拆线器；拆缝器

Exercises

List the types and functions of sewing machine.

导入：如何用英文表述服装设备？

第十章　服装设备

常用的服装设备工具有：测量工具、描图工具、熨烫工具、缝纫工具、裁剪工具。

第一节　测量工具

最常用的测量工具是卷尺：表面光滑，标有厘米和英寸。直尺是由透明的塑料制成，标有厘米和英寸。有2.5厘米（1英寸）宽、15.2厘米（6英寸）长和5.1厘米（2英寸）宽、45.7厘米（18英寸）长两种规格。码尺有91.4厘米（36英寸）和114.3厘米（45英寸）两种规格，由金属或者木料制成。裙摆标示器：裙摆标示器用于精确地标示裙摆的长度，可与画粉或者大头针一起用。

第二节　描图工具

滚轮将样板上的线条转移到面料上，并避免勾纱。描图纸用来将纸样上的线条转移到面料上。画粉是由蜡或者石头制成。蜡制画粉用于羊毛织物，其他面料可用石制画粉。曲线板是服装中绘制各部位弧线的工具。

第三节　熨烫工具

熨烫工具包括蒸气两用熨斗、挂烫机熨烫板、烫袖板、小烫板和烫凳。

第四节　缝纫工具

缝纫工具包括各类缝纫机，例如，锁式线迹缝纫机、锯齿形锁缝缝纫机、包缝机、安全线迹包缝机和锁眼机。压脚也是缝纫机的部件之一。压脚分为嵌线压脚、滚边压脚和打褶压脚。嵌线压脚分为左右两个压脚，用来平衡两边，用于缝嵌线和装拉链的暗针缝。滚边压脚用来将斜布条滚边成带。打褶压脚用于均匀的缝制永久性的褶。

此外，缝纫工具还包括手缝针、缝纫机针、大头针、顶针和蜂蜡。

第五节 裁剪工具

常用的服装裁剪工具有弯柄裁剪刀、纱线剪刀、花齿剪和拆线器（拆缝器）。

Garment Producing and Packaging
服装生产和包装

Chapter 11　Production Order
第十一章　生产通知单

项目名称：生产通知单

项目内容：1. 上装的生产通知单

　　　　　2. 下装的生产通知单

教学学时：6课时

教学目的：让学生了解生产通知单的格式和要点；并掌握生产通知单的翻译，培养学生服装外贸单据的翻译能力。

教学方式：由教师通过课前导入任务引出本章内容。分析讲解生产通知单的制作格式和要点，重点讲解生产通知单的翻译。归纳本章学习难点和重点，组织学生进行生产通知单的实操练习。

教学要求：（1）掌握生产通知单中的专业术语。

　　　　　（2）熟悉生产通知单中的格式、要点。

　　　　　（3）在外贸案例中熟练运用。

Chapter 11   Production Order

Part 1   Production Order of Top

Production order of top as Table 11-1 shown.

Table 11-1   Production Order of Top

HONDDA GARMENT CO., LTD. PRODUCTION ORDER

Page: 1

To: HUIDA	Date: 22-Aug-18
Style No.: 8HT4085WHY2	Style: MEN' S L/S SHIRT
Total Quantities: 1, 140 PCS.	Delivery Date: 1-Oct-18
Garment: No Wash	Lot NO.: 09003W

These production information sheets shall form integral parts of manufacturing contract no.

SPEC.: Sample must be approved before production

PART	S	M	L	XL
Center Back Length	30 1/2″❶	31	31 1/2″	32
Across Shoulder	21 1/4″	22 1/4″	231/4″	24 1/4″
Sleeve Length (From CB Neck)	33 1/2″	34 1/2″	35 1/2″	36 1/2″
Collar Length (From Button to Buttonhole)	15 1/2″	16 1/2″	17 1/2″	18 1/2″
Collar Spread	15 1/2″	16 1/2″	17 1/2″	18 1/2″
Collar Height	1″	1″	1″	1″
Mock Collar Height	1 1/2″	1 1/2″	1 1/2″	1 1/2″
Sleeve Placket Length	5 1/2″	5 1/2″	5 1/2″	5 1/2″
Sleeve Placket Width	1″	1″	1″	1″
Armhole Curved	11″	11 1/2″	12″	12 1/2″
Cuff Opening	8 1/2″	9″	9 1/2″	10″
Cuff Height	2 1/2″	2 1/2″	2 1/2″	2 1/2″
Chest(1″below armhole)	1 1/4″	1 1/4″	1 1/4″	1 1/4″
Waist	32″	33″	34″	35″
Sweep	33″	34″	35″	36″
Back Yoke Height(From CB Neck)	4 1/2″	4 1/2″	4 1/2″	4 1/2″
Front Placket Width	1 1/2″	1 1/2″	1 1/2″	1 1/2″
Welt Height	—	—	—	—
Welt Length	—	—	—	—

❶ 1″≈2.54cm.

Continued

STYLE   DETAIL

Page： 2

PLEASE PRODUCE ACCORDING TO APPROVAL SAMPLES

 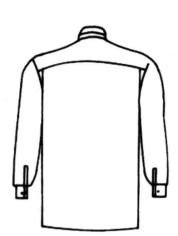

PRE-PRODUCTION SAMPLE: 2 PCS. IN SIZE M (AFTER RECEIVED FABRIC, BEFORE CUT)

SALESMAN SAMPLE: 3 PCS. IN SIZE L (TO MAKE AFTER SUBMITTED PRE-PRODUCTION SAMPLE)

SHIPMENT SAMPLE: 2 PCS. IN SIZE L (TO MAKE AFTER SUBMITTED PRE-PRODUCTION SAMPLE)

BREAKDOWN

PO. NO.	STYLE NO.	SIZE	S	M	L	XL	TOTAL
		RATIO	2	4	4	2	12
8593TA	8HT4085WHY2	—	190	380	380	190	1,140PCS.

FABRIC

PATTERN: 8JSO16   COLOR:WHITE

COMPOSITION: POLYESTER/RAYON WOVEN BENGALINE

WIDTH:57/58″   QUANTITIES: 22 YDS./DOZ.

REMARK: FABRIC SAMPLE:

Continued

		Page: 3
MATERIALS DETAILS		QUANTITY
THREAD	#402 MATCH THREAD (FACTORY ORDER)	—
INTERLINING PLACEMENT	# 3530F FUSIBLE INTERLINING (FACTORY ORDER) COLLAR, CUFF	/DOZ.
BUTTON PLACEMENT	18L❶ #JB-022 BK PLASTIC BUTTONS COLLAR (2 PCS.), CUFF(2 PCS.)	4PCS.
PLACEMENT	20L 4 HOLES MATCHING PLASTIC BUTTONS FRONT FLY	6PCS.
MAIN LABEL PLACEMENT	"EXCETERA" MATCHING LABEL COLOR THREAD, SEW ALL SIDES OF LABEL AT CENTRE BACK YOKE, BELOW THE NECK LINE 1/2″	1PC.
SIZE & COMPOSITION LABEL PLACEMENT	MATCHING MAIN LABEL COLOR INSERT AT THE BOTTOM OF MAIN LABEL	1PC.
WASHING LABEL PLACEMENT	MATCHING MAIN LABEL COLOR SEW AT RIGHT SIDESEAM, 2″ ABOVE HEM	1PC.
SIZE/STYLE NO. STICKER PLACEMENT	TRANSPARENT GROUND, BLACK LETTERS STICK AT WRONG SIDE OF HANG TAG	1PC.
HANG TAG PLACEMENT	"EXCETERA" AS SKETCH	1PC.
PRICE TICKET PLACEMENT	AS SKETCH	1PC.
HANGER	#170-78T	1PC.
PIN PLACEMENT	NO	0
CLIP PLACEMENT	NO	0
TISSUE PAPER PLACEMENT	NO	0
POLYBAG PLACEMENT	NO	0
INNER BOX PLACEMENT	(FACTORY ORDER)	6PCS./BOX 190BOXES
CARTON PLACEMENT	(FACTORY ORDER)	12PCS./CTN. 95CTNS.
TAPE STICKER PLACEMENT	WHITE GROUND, BLACK LETTERS STICK ON THE SHIPPING MARK, EACH 1″ FROM BOTTOM AND SIDE EDGE	2PCS./BOX 190PCS.

❶ 1L=0.633mm.

Continued

PACKING   DETAILS

Page： 4

(1) INNER BOX

6 PCS. ACCORDING TO THE RATIO

SIZE: S M L XL

QUANTITY: 1   2   2   1   = 6 PCS.

(2) EXPORT CARTON

ACCORDING TO THE RATIO 12 PCS. (2 INNER BOXES)

SIZE: S M L XL

QUANTITY: 2   4   4   2   = 12 PCS.

REMARKS:

SHIPPING MARK

INNER BOXES

TOTAL: 190 BOXES

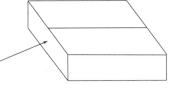

| STYTLE# 8HT4085WHY2 |
| LOT# 8593TA |

SIZE	S	M	L	XL	TOTAL
PCS.	1	2	2	1	6

EXPORT CARTON

TOTAL: 95 CTNS.

SHIPPING MARK	SIDE MARK
STYTLE# 8HT4085WHY2	STYTLE# 8HT4085WHY2
EXCETERAA	LOT# 8593TA
LOT# 8593TA	N.W.
1 DZ.	G.W.
MI MI	MEAS.
MADE IN CHINA	SIZE： S M L XL
C/N	QTY： 2 4 4 2 =12PCS.
AD DATE	COLOR： WHITE

TAPE STICKER

EACH 1″ FROM BOTTOM AND SIDE EDGE

Continued

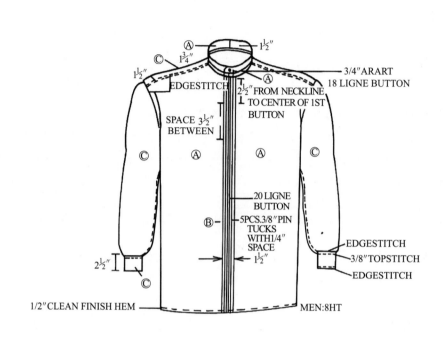

Ⓐ 1½″
Ⓒ 1¾″
1½″
EDGESTITCH
3/4″ARART
18 LIGNE BUTTON
2½″ FROM NECKLINE TO CENTER OF 1ST BUTTON
SPACE 3½″ BETWEEN
Ⓒ Ⓐ Ⓐ Ⓒ
20 LIGNE BUTTON
Ⓑ 5PCS.3/8″PIN TUCKS WITH1/4″ SPACE
1½″
EDGESTITCH
3/8″TOPSTITCH
EDGESTITCH
2½″
Ⓒ
1/2″CLEAN FINISH HEM
MEN:8HT

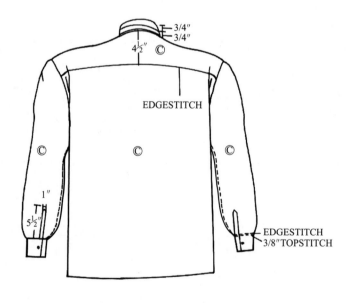

3/4″
3/4″
4½″ Ⓒ
EDGESTITCH
Ⓒ Ⓒ Ⓒ
1″
5½″
EDGESTITCH
3/8″TOPSTITCH

Part 2　Production Order of Bottom

Production order of bottom as Table 11-2 shown.

Table 11-2　Production Order of Bottom

HUAFENG GARMENT CO., LTD					
PRODUCTION ORDER					

Page: 1

STYLE NO.:_____　MERCHANDISER:_____　DATE:_____

FACTORY:_____　BUYER:_____

DELIVERY:_____　P.O. NBR.:_____

1.DESCRIPTION:			CONTRACT NO.:		QTY.:
FRONT			BACK		

2.FABRIC SAMPLE

(A)	(B)	(C)	(D)	(E)	(F)

3.LINING SAMPLE

(A)	(B)	(C)	(D)	(E)	(F)

4.SAMPLING

APPROVAL SAMPLE: SIZE_____QTY._____(MUST BE SUBMITTED BEFORE 9/15)

PRE-PRODUCTION SAMPLE: SIZE_____QTY._____(MUST BE SUBMITTED BEFORE PRODUCTION)

SHIPMENT SAMPLE: SIZE_____QTY._____(MUST BE SUBMITTED BEFORE 9/30)

Continued

COLOR & SIZE BREAKDOWN

Page: 2

STYLE: _____ ITEM: _____ SIZE: _____

SIZE			S	M	L		TTL(pcs.)
COLOR							
		Thread Color No.					
1)	Baby Blue	JIMTAI#562		200	300	200	=	700
2)	Cream	JIMTAI#310		100	200	200	=	500
							_____
							1200

Continued

PART		S	M	L		
chest width						
center back length						
shoulder width						
armhole girth						
biceps						
sleeve length						
cuff width						
cuff high						
bottom width						
neck circumference						
collar high						
front neck depth						
back neck depth						
pocket						
waist (relaxed)						
waist (stretched)						
hight hip						
low hip						
front rise						
back rise						
thigh						
bottom width						
outseam length						
inseam length						
waistband width						

SIZE SPECIFICATION

STYLE: _____ (Unit:cm)   Page: 3

Continued

STYLE: _____							Page: 4
ASSORTMENT SHEET							
COLOR		**S**	**M**	**L**			**PC./CART**	**TOTAL QTY.**
Baby Blue		2	3	2		=	7	
Cream		1	2	2		=	5	
						12pcs./group	
						NUMBER OF CARTON:17	
SHIPMENT							
QUANTITY		1, 200PCS.					

PACKING:

1. One piece into a polybag.

2. Two groups （total 24pcs.） into an inner box.

3. Three inner boxes into an export carton （total 72pcs.）.

Continued

BASIC ACCESSORIES

STYLE:_____   Page: 5

Accessories	QTY.	Kind	Placement	Remarks	Supplier
zippeer					
button					
shoulder pad					
snap					
interlining					
lining					
drawstring					
Label	QTY.	Kind	Placement	Remarks	Supplier
main label					
washing label					
size label					
origin label					
content label					
hang tag					
sticker					
price ticket					
Packing	QTY	Kind	Size	Remarks	Supplier
poly bag					
inner box					
export carton					
hanger					
pin					
tissue paper					

Continued

SHIPPING MARK

Page: 6

STYLE: _____   DATE: _____

SHIPPING MARK

S.M.E.I. SRL
ORDER NO.: 92/02676
STYLE NO.: 3750.01
QTY: 72 PCS .
CTN NO.:
MADE IN CHINA

SIDE MARK

G.W. :   KGS

N.W. :   KGS

MEAS. :   ×   ×   CM

REMARKS:

New Words and Expressions

1. P/O（production order）　生产制造单
2. across shoulder　肩宽
3. collar spread　（衣领）两领尖间距离
4. mock　假的；模拟的
5. placket　开口；开襟；袖衩
6. sweep　下摆（裙摆）；（衣裙等的）拖拽
7. welt　贴边；嵌线；嵌条
8. pattern　花样；式样；纸样
9. fusible interlining　热熔衬；黏合衬
10. front fly　前门襟；前筒
11. composition label　成分标识（唛）
12. sticker　贴纸
13. tape sticker　条码贴纸
14. ground　底子；板面
15. shipping mark　箱标识（唛）
16. Ligne　莱尼（纽扣规格：1 莱尼=0.633mm）
17. pin tuck　细褶；狭裥；细绉
18. edgestitch　缝边线迹；压边线
19. topstitch　正面线迹；面缝线迹；压面线
20. merchandiser　采购员；供销员
21. biceps　臂围
22. armhole width　袖窿宽；挂肩
23. thigh　大腿；大腿围；股
24. outseam length　裤长；外缝长
25. origin label　产地唛
26. hanger　衣架
27. pin　大头针；胶针

Exercises

1. Design a production order form in English by yourself.

2. Translate the production order (including two pages) below.

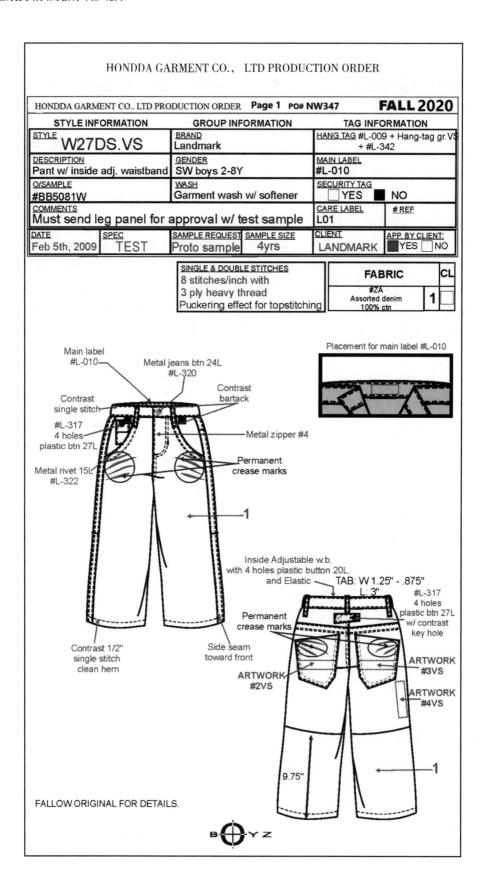

HONDDA GARMENT CO., LTD PRODUCTION ORDER

HONDDA GARMENT CO., LTD PRODUCTION ORDER	Page 1 PO# NW347	FALL 2020

STYLE INFORMATION	GROUP INFORMATION	TAG INFORMATION
STYLE **W27DS.VS**	BRAND Landmark	HANG TAG #L-009 + Hang-tag gr. VS + #L-342
DESCRIPTION Pant w/ inside adj. waistband	GENDER SW boys 2-8Y	MAIN LABEL #L-010
O/SAMPLE #BB5081W	WASH Garment wash w/ softener	SECURITY TAG ☐ YES ■ NO
COMMENTS Must send leg panel for approval w/ test sample		CARE LABEL L01 # REF

DATE Feb 5th, 2009	SPEC TEST	SAMPLE REQUEST Proto sample	SAMPLE SIZE 4yrs	CLIENT LANDMARK	APP. BY CLIENT: ■ YES ☐ NO

SINGLE & DOUBLE STITCHES 8 stitches/inch with 3 ply heavy thread Puckering effect for topstitching	FABRIC #ZA Assorted denim 100% ctn	CL 1 ☐

Main label #L-010

Metal jeans btn 24L #L-320

Contrast bartack

Contrast single stitch

#L-317 4 holes plastic btn 27L

Metal zipper #4

Permanent crease marks

Metal rivet 15L #L-322

1

Placement for main label #L-010

Inside Adjustable w.b. with 4 holes plastic button 20L. and Elastic

TAB: W 1.25" - .875" L: 3"

#L-317 4 holes plastic btn 27L w/ contrast key hole

Permanent crease marks

Contrast 1/2" single stitch clean hem

Side seam toward front

ARTWORK #2VS

ARTWORK #3VS

ARTWORK #4VS

9.75"

1

FALLOW ORIGINAL FOR DETAILS.

B⊕YZ

HONDDA GARMENT CO., LTD PRODUCTION ORDER **Page 2**   **PO# NW347**   **FALL 2020**

STYLE INFORMATION	GROUP INFORMATION	TAG INFORMATION
STYLE **W27DS.VS**	BRAND Landmark	HANG TAG #L-009 + Hang-tag gr.VS + #L-342
DESCRIPTION Pant w/ inside adj. waistband	GENDER SW boys 2-8Y	MAIN LABEL #L-010
O/SAMPLE #BB5081W	WASH Garment wash w/ softener	SECURITY TAG ☐ YES ■ NO
COMMENTS Must send leg panel for approval w/ test sample		CARE LABEL L01 #REF
DATE Feb 5th, 2009	SPEC TEST	SAMPLE REQUEST Proto sample / SAMPLE SIZE 4yrs	CLIENT LANDMARK	APP. BY CLIENT: ■ YES ☐ NO

These specs are in inches and circumference

W27DS.VS Boys pant
Denim, 100% cotton

		2	4	6	8	tol +/-
1	WAIST CIRC (FLAT)	20 1/2	22	24	26	1/2
2	HIPS CIRC.(FLAT)	25 1/2	27 1/2	29 1/2	31 1/2	1/2
3	HIPS LEVEL FROM CROTCH	1 3/4	2	2 1/4	2 1/2	0
4	THIGH CIRC (FLAT) meas. 1" below crotch	15 1/2	16 1/2	17 1/2	18 1/2	1/4
5	FRONT RISE incl. w/b	6 7/8	7 5/8	8 3/8	9 1/8	1/4
6	BACK RISE incl. w/b	9 3/4	10 1/2	11 1/4	12	1/4
7	INSEAM REG.LEG (LONG PANT)	14	17	20	23	1/2
8	OUTSEAM	21	24 1/2	28	31 1/2	1/2
9	KNEE LEVEL FROM CROTCH	6	7 1/2	9	10 1/2	0
10	KNEE CIRC meas. REG. LEG (LONG PANT)	13 1/4	14	14 3/4	15 1/2	1/4
11	LEG OPENING CIRC REG. LEG (LONG PANT)	13	13 1/2	14	14 1/2	1/4
12	BOTTOM HEM (1-needle) clean finish	5/8	5/8	5/8	5/8	0
13	FRONT PKT OPENING **ALONG WB**	2 1/8	2 3/8	2 5/8	2 7/8	1/8
14	FRONT PKT OPENING **ALONG SIDE**	4 1/8	4 1/8	4 1/8	4 1/8	1/8
15	FRONT PKT BAG WIDTH	4 1/4	4 1/2	4 3/4	5	1/8
16	FRONT PKT BAG HEIGHT	5 1/2	5 3/4	6	6 1/4	1/8
17	W/BAND HEIGHT	1 1/2	1 1/2	1 1/2	1 1/2	1/8
18	FLY **ZIPPER LENGTH** (WB not inc)	2 1/2	3	3 1/2	4	1/8
19	FLY **WIDTH**	1 3/8	1 3/8	1 3/8	1 3/8	1/8
20	LOOPS W X L	1/2 x 2 1/4	3/8 x 2 3/8	1/2 x 2 1/4	1/2 x 2 1/4	1/8
21	BACK YOKE HEIGHT (CB - side)	2 - 1 1/4	2 - 1 1/8	2 - 1 1/4	2 - 1 1/4	1/8
22	BACK POCKET WIDTH	4 1/2	4 3/4	5	5 1/4	1/8
23	BACK POCKET HEIGHT	5 5/8	5 7/8	6 1/8	6 3/8	1/8
24	FRONT WATCH POCKET WIDTH	1 1/2	1 3/4	2	2 1/4	1/8
25	FRONT WATCH POCKET HEIGHT	2 1/4	2 1/2	2 3/4	3	1/8

*PROTOSAMPLE
Please send protosample with actual color / accesories & fabric. To put artwork on protosample.
If fabric and/or accesories are not ready, pls send proto in the closest color and fabric.
Please indicate on the hang tag what is actual and what is available.

Please send protosample in size 4Y to us (Canada)
Please send protosample only in size 4Y to Landmark.
We will do fitting and styling approval on protosample and give you final specs.
After protosample is approved, please make corrections and send Pre-production samples
in the same sizes

Fabric # W-26 ,Refer to denim card

Instructions (PANT):
-to have bar tack at all pockets opening, loops and bottom fly.
-To have bar tack at inside crotch
-Zipper end to be tack
-Pocket bag to be in off white pocketing with self entry facing
-Button hole to be key hole style
-Shank button to use proper size & height, to be solid & properly install.
-Inside adj. Elastic s/b tacked at both ends, underneath front loops
-TOPSTITCH: 8 stitches/inch. To use heavy 3 ply thread

参考译文

导入：在外贸中，如何理解和翻译服装生产通知单？

第十一章　生产通知单

第一节　上装的生产通知单

上装的生产通知单见表11-1。

表11-1　上装的生产通知单

宏达制衣有限公司生产通知单			
			第1页

给：晖达		日期：2018年8月22日	
款号：8HT4085WHY2		款式：男装长袖衬衫	
总数量：1140件		交货日期：2018年10月1日	
成衣：不洗水		批号：09003W	
这些生产资料单是加工合同的构成部分			
说明：生产前样板必须经审批			

部位	S	M	L	XL
后中长	30 1/2″	31	31 1/2″	32
肩宽	21 1/4″	22 1/4″	231/4″	24 1/4″
袖长（从后中度量）	33 1/2″	34 1/2″	35 1/2″	36 1/2″
领围（从纽扣量至扣眼）	15 1/2″	16 1/2″	17 1/2″	18 1/2″
领尖距	15 1/2″	16 1/2″	17 1/2″	18 1/2″
后领高	1″	1″	1″	1″
假领高	1 1/2″	1 1/2″	1 1/2″	1 1/2″
袖襟长	5 1/2″	5 1/2″	5 1/2″	5 1/2″
袖襟宽	1″	1″	1″	1″
袖窿（弯度）	11″	11 1/2″	12″	12 1/2″
袖口宽	8 1/2″	9″	9 1/2″	10″
袖口高	2 1/2″	2 1/2″	2 1/2″	2 1/2″
胸围（袖窿下1英寸处量度）	1 1/4″	1 1/4″	1 1/4″	1 1/4″
腰围	32″	33″	34″	35″
底摆	33″	34″	35″	36″
后育克高（从后中量）	4 1/2″	4 1/2″	4 1/2″	4 1/2″
前门襟宽	1 1/2″	1 1/2″	1 1/2″	1 1/2″
嵌线袋高	—	—	—	—
嵌线袋长	—	—	—	—

续表

款式细节

第 2 页

请根据确认样板生产

 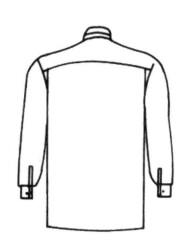

产前样：中码 2件　　（收布后，开裁前）

展销样：大码 3件　　（大货出运前即做）

（装）船样：大码 2件（大货出运前即做）

数量分配

制单号	款号	尺码	S	M	L	XL	共计
		比率	2	4	4	2	12
8593TA	8HT4085WHY2	数量	190	380	380	190	1140件

布料

花样：8JSO165	颜色：白色
成分：涤纶/黏胶　机织罗缎	
幅宽：57/58英寸	用量：22码/打

备注：	布料样板：

物料细节			数量
线		402# 顺色线　　　　　　　　　　　　　　　（工厂订）	—
衬	位置	#3530F 黏合衬　　　　　　　　　　　　　　（工厂订） 领，袖头	/打
纽扣	位置	18L #JB-022 黑色胶纽 领（2粒），袖克夫（2粒）	4粒
	位置	20L 4孔顺色胶纽 前门襟	6粒
主唛/商标	位置	"EXCETERA" 商标顺色线，四边车缝于后中育克内，领围线向下1/2英寸	1个
尺码和成分唛	位置	主唛顺色线 嵌插车缝于主唛底	1个
洗水唛	位置	主唛顺色线 车缝于右侧缝，底边向上2英寸	1个
尺码/款号贴纸	位置	透明底，黑字 贴在吊牌背面	1个
吊牌	位置	"EXCETERA" 如图示	1个
价格牌	位置	如图示	1个
衣架		#170-78T	1个
大头针	位置	无	0
塑料夹（胶夹）	位置	无	0
拷贝（包装）纸	位置	无	0
塑料（胶）袋	位置	无	0
内盒	位置	（工厂订）	6件/每内盒 190盒
外箱	位置	（工厂订）	12件/外箱 95箱
条码贴纸	位置	白底黑字 贴在外箱正唛处，距底边及箱侧各1英寸	2个/每箱190箱

第 3 页

续表

包装要求

（1）内盒

6件 按比例

尺码： S M L XL

数量： 1 2 2 1 ＝6件

（2）外箱

按比例 12件（2内盒）

尺码： S M L XL

数量： 2 4 4 2 ＝12件

备注：

箱唛

内盒

共：190盒

款号： 8HT4085WHY2

批号： 8593TA

尺码	S	M	L	XL	共计
件数	1	2	2	1	6

外箱

共：95箱

箱唛

款号：8HT4085WHY2

EXCETERA

批号：8593TA

1 打

MI MI

中国制造

箱号

核准日期

侧唛

款号：8HT4085WHY2

批号：8593TA

净重：

毛重：

规格：

尺码： S M L XL

数量： 2 4 4 2 ＝12件

颜色：白色

条码贴纸

距底边及箱侧各1英寸

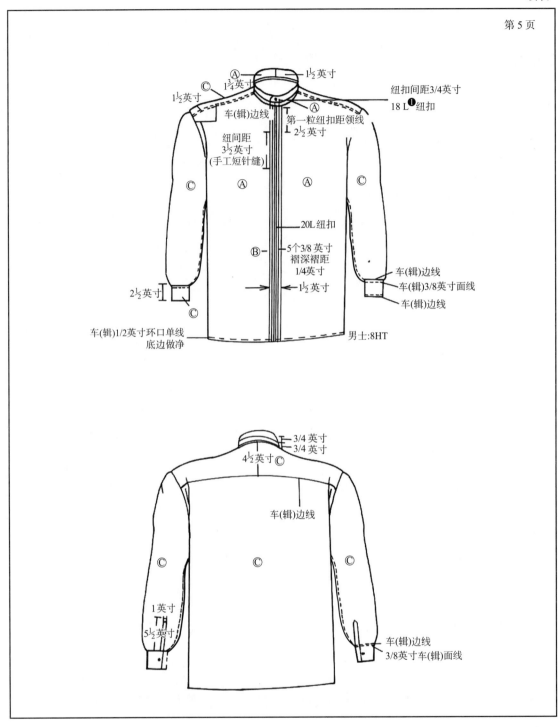

纽扣间距3/4英寸
18 L❶纽扣

1½英寸

Ⓐ
1¼英寸
Ⓒ
1½英寸
车(辑)边线

第一粒纽扣距领线
1½英寸
Ⓐ

纽间距
3½英寸
(手工短针缝)
Ⓒ
Ⓐ
Ⓐ
Ⓒ

20L纽扣

Ⓑ
5个3/8英寸
褶深褶距
1/4英寸

2½英寸
车(辑)边线
车(辑)3/8英寸面线
车(辑)边线

1½英寸
Ⓒ

车(辑)1/2英寸环口单线
底边做净
男士:8HT

3/4 英寸
3/4 英寸
4½英寸Ⓒ

车(辑)边线

Ⓒ
Ⓒ
Ⓒ

1英寸
5½英寸

车(辑)边线
3/8英寸车(辑)面线

---

❶ 1L（莱尼）=0.633mm.

第二节　下装的生产通知单

下装的生产通知单见表11-2。

表11-2　下装的生产通知单

华丰服装有限公司

生产通知单

第1页

款号：＿＿＿＿＿＿＿　跟单员：＿＿＿＿＿＿＿　日期：＿＿＿＿＿

工厂/口岸：＿＿＿＿＿＿＿＿＿＿　买方：＿＿＿＿＿＿＿

交货日期：＿＿＿＿＿＿＿＿＿　订单号：＿＿＿＿＿＿＿

1.款式：	合同编号：	数量：

前幅　　　　　　　　　　　后幅

2.面料样板					
（A）	（B）	（C）	（D）	（E）	（F）

3.里料样板					
（A）	（B）	（C）	（D）	（E）	（F）

4.样板

确认样＿＿＿码＿＿＿件（须于9月25日前送抵我方确认）

产前样＿＿＿码＿＿＿件（须于大货生产前送抵我方确认）

船样＿＿＿码＿＿＿件（须于9月30日前送抵我方确认）

续表

颜色及尺码搭配表

第2页

款号：＿＿＿＿＿＿＿＿＿＿ 品种：＿＿＿＿＿＿＿＿＿＿ 尺码：＿＿＿＿＿＿＿＿＿＿

尺码				S	M	L		共（件）
颜色								
		缝线色号						
1）	浅蓝	金泰#562		200	300	200	=	700
2）	米色	金泰#310		100	200	200	=	500
								————
								1200

部位		S	M	L		
胸围						
后中长						
肩宽						
袖窿围						
臂围						
袖长						
袖口长						
袖口高						
下摆宽						
领围						
后领高						
前领深						
后领深						
口袋						
腰围（放松量）						
腰围（拉伸量）						
上臀围						
下臀围						
前裆弧长（前浪）						
后裆弧长（后浪）						
大腿围（1/2）						
脚口宽						
外长						
内长						
腰头高（宽）						

尺寸表

款号：＿＿＿＿＿＿＿＿（单位：cm）　　　　　　　　　　　　　　第3页

续表

| 款号： _____ | | | | | | | 第4页 | |

入箱搭配								
颜色		小号	中号	大号			件/箱	总件数
浅蓝		2	3	2		=	7	
米色		1	2	2		=	5	
							12件/组	
							箱数：17	
交期								
总数		1200件						

包装要求：

1. 每件入1塑料（胶）袋。

2. 每两组（共24件）入1内盒。

3. 3内盒入1外箱（共72件）。

辅料

款号：_____ 第5页

辅料	数量	品别	位置	备注	供应商
拉链					
纽扣					
肩垫					
按扣（揿纽）					
衬料					
里料					
束带（绳子）					

商标	数量	品别	位置	备注	供应商
主标					
洗涤标					
尺码标					
产地标					
成分标					
吊牌					
贴纸					
价格牌					

包装要求	数量	品别	规格大小	备注	供应商
塑料（胶）袋					
内盒					
外箱					
衣架					
别针（胶针）					
包装纸（拷贝纸）					

续表

外箱唛

第6页

款号：_____

日期：_____

正唛

S.M.E.I.   SRL❶(客户公司名)
订单号：92/02676
款号：3750.01
数量：72件
箱号：
中国制造

侧唛

毛重：   千克

净重：   千克

规格：   ×   ×   厘米

备注：

---

❶ SRL是意大利语，中文意思为（股份）责任有限公司，相当于英文的LTD。

Garment Producing and Packaging
服装生产和包装

Chapter 12 Quality Control and Inspection Report
第十二章　质量控制与检验报告

项目名称： 质量控制与检验报告

项目内容： 1. 检验方法

2. 质量标准

3. 测量指南

4. 服装疵点的表述

5. 检验报告

教学学时： 6课时

教学目的： 让学生了解制衣业常用的质检方法、质量要求、服装疵点的表述和检验报告表的填写；掌握服装测量部位、质量标准、服装疵点等英文表述。

教学方式： 由教师通过课前导入任务引出本章内容。分析讲解质检方法、质量标准、测量指南和检验报告表的格式和要点，重点讲解服装质检疵点的英文表述。通过拓展视频，深入学习服装质检。归纳本章学习难点和重点，组织学生进行检验报告表实操练习。

教学要求： （1）掌握服装疵点的英文表述。

（2）熟悉检验报告表的格式、表中要点的格式翻译。

（3）在外贸案例中熟练运用。

Chapter 12   Quality Control and Inspection Report

Part 1   Methods of Inspection

The inspection activity of itself adds nothing to the product. It is only a means by which information is obtained as a basis for further action.

Effective inspection involves six main functions:

(1) Select the item for inspection, may be a component, subassembly or finished item.

(2) Define the standard required, specifications, samples, sketches, diagrams.

(3) Examine the item critically, by observation, measurement or quantitative evaluation testing.

(4) Compare the observation with the defined standard.

(5) Decide acceptability or otherwise.

(6) Act on the item according to the decision, return for correction, go forward to stock or to next operation etc.

Each of these activities must be completed if inspection is to be effective. Inspection is "the obtaining of information" about the item or product. Whether the information is useful to the quality control will determine how effective the inspection is.

Various methods of carrying out the "inspection" function are in use in the garment industry. The selection of an appropriate method will be governed by specific circumstances and aims. Some methods are more suitable for the prevention of errors or faults, some for controlling the quality of production, and others for determining the quality level of the finished product.

Inspection methods in common use in the garment industry are:

(1) Final inspection.

(2) In process inspection.

(3) Centralized inspection.

(4) Patrol inspection.

(5) Sampling inspection.

(6) Directed sampling inspection.

(7) Pre-production inspection.

(8) Total quality control (quality assurance).

Most garment factories use one or a combination of these methods.

Part 2   Quality Standards

1   Introduction

In garment industry, "Quality" refers to the standard of work and output. A manufacturer with a good name or well-known label will never lower the quality of it's goods to reach the price, profit and output desired, but will seek to use new and better methods of making more efficient usage of factory space, personnel and machinery.

2   Manufacturing Specifications

In practice, for individual manufacturing specifications, a specification key should be drawn up to cover all aspects of work in the sewing room before producing, and this will act as an outline and a check that all items have been covered in individual operations. It is convenient to consider sewing room operations under the following headings and some examples.

(1) Positioning of components:

①Seam allowance.

②Position start/ finish.

③Backstitch.

④Bartack.

(2) Matching details:

①Matching stripes and checks.

②Positioning of ends/ edges/ notches/ seams.

③Location of tapes/ label/ hanger.

④Fullness/ easing.

(3) Stitching detail:

①Type.

②Stitch density.

③Skipped/ broken/ missed stitches.

④Tension appearance.

⑤Thread color.

⑥Thread ends length.

⑦Damaged thread.

(4) Appearance of completed process:

①Fullness.

②Puckering.

③Attached thread.

④Balancing left & right.

⑤Stripes and checks.

⑥Pleating.

⑦Distortion.

⑧Material.

⑨Face & wrong side.

⑩Color shading.

⑪Soiled.

⑫Edges/ corners.

(5) Pressing Operations:

①Fullness.

②Flatness.

③Pleating.

④Distortion.

⑤Marks.

⑥Fused interlining.

(6) Marking/ cutting/ notching detail:

①Shape.

②Position.

③Template.

(7) Customer Requirements:

①Additional special detail.

②Outside standard specifications.

Part 3　How to Measure with Guidelines

1　Tops

(1) Center back length: Lay the garment flat with the back of the garment facing you. Measure from the center of the back neck seam to the bottom of garment.

(2) Shoulder width: Lay the garment flat with the back of the garment facing you. Locate the shoulder points where the shoulder seams meet the top of armhole, measure straight across from shoulder point to shoulder point. When there is no natural shoulder seam, measure where the natural fold of the shoulder meets the top of the armhole.

(3) X Back: Lay the garment flat with the back of the garment facing you. Measure down from the center back neck seam at the specified point on the size chart, measure straight across from

armhole seam to armhole seam.

(4) Nape to Waist: Lay the garment flat with the back of the garment facing you. Measure down the back center of the garment from neck seam to waist seam.

(5) Sleeve length: Lay the garment flat with sleeve and shoulder free of wrinkles. Measure from the center of the back neck, across shoulder point and following the centre-fold of sleeve to bottom.

(6) Cuff: Lay the sleeve flat (sleeve closed for button styles). Measure straight across the bottom of the sleeve fold to fold.

(7) Chest/ Bust: Lay the garment flat. Measure straight across garment from side to side at the bottom of armholes. Measure down from armhole at side seam, on garments with fullness at bust.

(8) Hem: Lay the garment flat, and make sure that any pleat is fully extended. Measure at bottom of garment following the contour of hem.

(9) Back neck: Measure the amount of inches the garment is cut away from the natural neck.

(10) Armhole: Lay the garment flat, aligning front and back armhole seams and position it so that the armhole seam has no wrinkles. Measure from top to bottom of armhole along seam contour.

(11) Collar length: Undo all closures and lay the collar flat so that the inside of garment is facing you. Measure along center of the collar stand from the center of the button to the end of the buttonhole.

2　Bottoms

(1) Waist: Measure from side to side along the waist seam. Measure from side to side along the center of the waistband on bottoms.

(2) Hip: Lay the trousers flat, consider to open the pleats according to the specified on size chart. Eliminate all wrinkles at the seat parts. With the tape measure straight across from side seam to side seam.

(3) Thigh: Lay the trousers flat, and eliminate all wrinkles at the thigh placement. Measure straight across from the crotch point to the side seam with the tape.

(4) Knee: Lay the trousers leg flat, and measure at the knee point from fold to fold, parallel to the bottom of trousers.

(5) Bottom: Lay the bottom flat, measure straight across the bottom from fold to fold.

(6) In-leg: Lay the trousers flat with one of the legs, free of wrinkles, and measure from crotch point to the bottom of the leg.

(7) Out-leg: Lay the bottoms flat, and eliminate all wrinkles. Measure from the waistband to the bottom on the same side.

(8) Front rise: Lay the bottoms flat and front face to you, and then eliminate all wrinkles on top part. Measure from center waistband to crotch point at the front rise.

(9) Back rise: Lay the bottoms flat and back face to you, and then eliminate all wrinkles on top part. Measure from waistband to crotch point at the back rise.

Part 4　Expression of Garment Defects

1　Defect of Jeans(Table 12-1)

Table 12-1　Jeans Defects

No.	Description	No.	Description
1	Rough yarn	21	Uneven tension at pkt. topstitch
2	Uneven dyeing	22	Press stud insecure
3	Shaded parts	23	Uneven front fly width
4	Waistband stitch skipped	24	Front fly length too long
5	High/ low waistband ends	25	Waistband twisted
6	Waistband end not close to front fly	26	Wrong waistband height
7	Raw edge at body	27	Stitch per inch less than specified
8	Uneven belt-loop size	28	Incorrect shape of back pocket
9	Missing belt-loop	29	Twisted bottom
10	Pocket lining caught in bartack	30	Fraying the end of catching facing
11	Too short pocket bag	31	Exposed drill holes
12	Inseam skipped stitch	32	Uneven top-stitch at left fly
13	Under layer exposed	33	Belt-loop not as sample
14	Pocket setting misplaced	34	Missing bartack
15	Pocket edge not straight	35	Measurement out of tolerance
16	Uneven size of back yoke	36	Wrong care label
17	Too much seam allowance in setting pocket	37	The under crotch meet point do not matched
18	Improper layer shown out	38	Zipper cannot reach at top
19	Mismatched meet point of back yoke	39	Out-seam run off stitch
20	Pocket-bag not extend to front fly		

2　Defect of Shirt(Table 12-2)

Table 12-2　Shirt Defects

No.	Description	No.	Description
1	Collar edge asymmetric	8	Twisted placket
2	Uneven collar stand	9	Incorrect pocket size
3	Collar tip grinning stitch	10	Under layer appeared after collar top stitch
4	Under layer exposed	11	Defective buttons
5	Uneven tension at topstitch collar	12	Collar size incorrect
6	Placket bubbling	13	Seam broken of cuff edge
7	Collar bubbling by uneven pressing	14	Uneven of collar tip

Continued

No.	Description	No.	Description
15	Uneven length of front placket	30	Seam slippage at bottom hemming
16	Pleated at sleeve joining	31	Collar bubbling due to incorrect handling
17	Improperly pressed	32	Seam broken at collar stand end
18	Collar tip seam broken	33	Pleats or puckers outside seriously affecting appearance
19	Wrong edge margin	34	Button not sewn through all eyes
20	Uneven height in both sides of cuff	35	Button sewn on face down
21	Thread fused into collar	36	Buttonholes cut in opposite direction to specification
22	Buttonholes stitching too narrow	37	One or both collar stay(s) missing
23	Side edge of pkt. exposed beyond side edge of flap	38	Wrong type of seam
24	Uneven tension of top stitch cuff	39	High/Low pkts. & flaps
25	Uneven shape in collar fall handling	40	Seam broken in sleeve placket
26	Woven label 1/2″ or more off center	41	Fullness on the under ply of the cuff
27	Uneven distance of slv. pleat	42	Incorrectly folded edge
28	Cuff assembling to slv. not closed	43	Twist cuff
29	Collar assembling to body not closed		

Part 5   Inspection Report

This form is used to record the information about the checked garment. The quality controller or checker would bring with this form while they are inspecting the goods. All the defects would be found and marked. The report would be submitted to the merchandiser(merchandise manager) director to take further remedy action(Table 12-3 to Table 12-7).

Table 12-3 Inspection Report

☐IN-LINE ☐FINAL										
FACTORY NAME: _____					DATE: _____					
ORDER NO.: ____ DESCRIPTION: _____ QUANTITIES: ____										
GOODS NO.: ____ DELIVERY DATE: ____ LOT NO.: ____										
Meas.	Spec.	Meas.Diff. S	Spec.	Meas.Diff. M	Spec.	Meas.Diff. L			Item	Quality Audit
front length									fabric handle	
c.b.length									main label	
shoulder width									size label	
chest									washing label	

Continued

Meas.	Spec.	Meas.Diff. S		Spec.	Meas.Diff. M		Spec.	Meas.Diff. L		Item	Quality Audit
waist	73	+1	+2	77	√		83	-1		hang tag	
sleeve length										poly bag	
cuff width										inner box	
armhole										export carton	
hip	99	+1		104	√		109	-1	-2	finishing	
bottom width	35	√		36	√		37	+1		assortment	
collar width										shipping mark	
collar heigh											
bicept											
ffront rise											
back rise											
thigh											
trousers length	100	+1		101	√		101	+1			

ITEMS/DEFECTS:

REMEDY ACTION OF FACTORY:

Factory's Signature _____ Inspector's Signature _____

Table 12-4   HONGDA Trading Co., Ltd.

MEASUREMENT CHART (TOPS)

SELLER: _____ BUYER'S ORDER NO.: _____ CONTRACT NO.: _____ STYLE: _____

LOT NO.: _____ QUANTITY: _____ BY: _____ DATE: _____

Meas.	S			M			L			XL		
	Spec.	Meas.	Diff.	Spec.	Meas.	Diff.	Spec.	Meas.	Diff.	Spec .	Meas.	Diff.
front length												
center back length												
shoulder width												
chest												
waist												
sleeve length												
cuff/sleeve opening												
armhole												
biceps												
bottom width												

Continued

Meas.	S			M			L			XL		
	Spec.	Meas.	Diff.	Spec.	Meas.	Diff.	Spec.	Meas.	Diff.	Spec .	Meas.	Diff.
collar width												
collar heigh												
front neck deep												
back neck deep												
pocket												
hood												

Table 12-5   HONGDA Trading Co., Ltd.

MEASUREMENT CHART (BOTTOMS)

SELLER: _____   BUYER'S ORDER NO.: _____   CONTRACT NO.: _____   STYLE: _____

LOT NO.: _____   QUANTITY: _____   BY: _____   DATE: _____

Meas.	S			M			L			XL		
	Spec.	Meas.	Diff.	Spec.	Meas.	Diff.	Spec.	Meas.	Diff.	Spec .	Meas.	Diff.
waist (relaxed)												
waist (stretched)												
high hip												
low hip												
thigh												
knee												
bottom width												
front rise												
back rise												
inseam length												
outseam length												
waistband width												
zipper length												

Table 12-6 Quality Control Department

IN-LINE CHECK REPORT

SELLER: _____ PRODUCT TYPE: _____ LOT NO.: _____ O/N: _____

SHIPMENT: _____ QUANTITY: _____ DATE: _____ BY: _____

(Ⅰ) Workmanship: findings and recommendations for production improvement

Items/Defects	Quantities	
	Inspected	Defective

(Ⅱ) Fit: (a) no wash/garment washed/stone washed/garment dyed (b) before/after wash (c) before/after pressing

Meas.	Spec.	Spec.	Spec.	Spec.	Spec.

(Ⅲ) Production status:

cut:

sew:

finish:

pack:

remark:

(Ⅳ) Acknowledged by Seller

Table 12-7 Defect Checklist in Process Inspection Report

Style No.: _____　　　　　　　　Page No.: _____

Fabric Defects	Major	Minor	Sewing Defects	Major	Minor
Holes			Open Seams		
Soiling			Weak Seams		
Flaws			Raw Edges		
Pilling			Puckering		
Uneven Dyeing			Wavy Stitches		
Burn Marks			Skip Stitches		
Bar			Broken Stitches		
			Incorrect Linking		

Garment Defects	Major	Minor	Garment Defects	Major	Minor
Fabric Color Mismatch			Pressing Defects		
Component Color Mismatch			Uneven Hem		
Defective Snaps			Misalliance Parts		
Defective Zippers			Missing Parts		
Exposed Zipper Tape			Uneven Plaids		
Excessive Thread Ends					
Loose Buttons					
Defective Buttonholes					

No. Pieces Accepted: _____　　No. Pieces Inspected: _____

Acceptance Level: _____

No. Pieces Rejected: _____　　　Minor Defects: _____

Date	Comments	Action to Be Taken by QC./MR.	Vendor's Initials
	PRODUCTION STATUS:		

The vendor is responsible for correcting all defects found during the inspection(s) and summarized in this report. However, the inspection does not relieve the vendor from its responsibility for defects found in the merchandise shipped.

Seller's Signature _____ Date _____ Inspector's Signature _____ Date _____

New Words and Expressions

1. product　产品
2. effective　有效的
3. involve　包含；包括
4. component　部分；成分
5. define　精确地解释(词等)的意义
6. sketch　草图；略图
7. diagram　图表
8. observation　观察；注意
9. evaluation　评价；估计
10. critically　批评地；吹毛求疵的
11. appropriate　适合的；适合于……的
12. govern　治理；控制
13. circumstance　环境；情势
14. patrol　巡逻；巡查
15. combination　联合；结合；组合
16. remedy　矫正；治疗
17. in-line　上线
18. handle　手感；手柄
19. workmanship　技艺；工艺；技巧
20. recommendation　建议；推荐；介绍
21. defect　缺点；短处；
22. acknowledge　公认；承认
23. soiling　弄脏；弄污；污迹
24. pilling　球粒疵
25. uneven dyeing　染色不匀
26. burn mark　烧斑
27. bar　横档疵
28. major　严重的
29. minor　轻度的；较小的
30. open seam　分开缝；开式缝
31. weak seam　弱缝口
32. raw edge　毛边
33. puckering　缩皱
34. wavy stitch　波形线迹

35. skip stitches　跳线
36. broken stitches　间断线迹
37. linking　套口；缝口
38. exposed　外露的；使暴露的
39. excessive　过多的；过度的；极端的
40. uneven hem　底边起绺
41. misalliance parts　鸳鸯片
42. uneven plaid　格子不匀
43. MR. (merchandiser)　跟单员
44. initials　姓名各字母的起首；签字
45. summarized　摘要的；概述的
46. quality standards　品质标准
47. garment industry　制衣业
48. output　产出
49. manufacturer　制造商
50. well-known label　知名品牌
51. machinery　机械；机器设备
52. manufacturing specification　制造规格
53. positioning of components　部件定位
54. seam allowance　缝份
55. backstitch　回针
56. stripes and checks　条格(布料)
57. notch　刀口；刀位
58. tapes　带条
59. fullness　放松量
60. easing　容位
61. stitch density　针迹密度
62. attached thread　粘线
63. distortion　变形
64. color shading　色差
65. template　样板
66. guidelines　指南
67. tops　上装
68. soiled　污渍

69. shoulder width　肩宽
70. X back　后背宽腰直
71. sleeve length　袖长
72. wrinkle　皱褶
73. chest/ bust　胸围
74. hem　下摆围
75. back neck　后领圈
76. contour　轮廓线
77. bottoms　下装
78. hips　坐围
79. eliminate　消除
80. parallel to　平行
81. in-leg　内长
82. out-leg　外长
83. garment defects　服装疵点

84. Jeans　牛仔裤
85. rough yarn　粗纱
86. waistband ends　裤腰嘴
87. tension　张力
88. press stud　工字纽扣
89. top stitch　缉面线
90. tolerance　宽余位
91. care label　洗水商标
92. fraying　散边；织物边缘磨损
93. asymmetric　不对称
94. grinning stitch　露齿针迹
95. placket　明筒
96. bubbling　起泡
97. seam broken　爆口
98. puckers　皱褶

Exercises

Translate the following emails, especially pay attention to the quality defects.

Date: Friday, December 28, 2019, 20:11:48
To: Angela@pub.huizhou.gd.cn
Cc: Bandy168@163.net
From: Barbara@winpap.co.uk
Subject: Quality Defects

Dear Angela,

　　There is a problem that needs urgent treatment. In the first shipment of children's pajamas, the diamond-ironing will fall off. This is absolutely not allowed. It must be ensured that the buttons, string beads and snaps must be firmly nailed to the pajamas. Inaccurate cotton bartacks - the bartacks did not set the position according to the process requirements.

B.RGDS,

Barbara

参考译文

第十二章　质量控制与检验报告

第一节　检验方法

在检验过程中，不可以对产品做任何的增添或更改，仅为今后的改进积累资料。

有效的检验方法必须包括以下六种功能：

（1）选择检验的项目，可以是衣服的一部分、半成品或是成品。

（2）说明需要的标准、规格、货板、设计图及图表。

（3）通过观察、测量或定量评估，对已选货品进行严格的检测。

（4）用确定的标准比较观察结果。

（5）决定接受与否。

（6）依照决定采取行动，送回原来部门做修改，送往货仓存货或送往下一程序处理。

若要得到有效的检验结果，必须完成以上每一步骤。总之，"检验"是为了取得关于该项目或产品的资料，所获信息是否对品质控制有用将决定该检验方法是否有效。

在制衣业有不同的检验方法，选择方法是基于不同的环境和目的，有些适合预防错误或疵点，有些是用来控制产品的质量，有些用于衡量制成品的质量水平。

以下是制衣业常用的检验方法：

（1）最终检验。

（2）工作程序中的检验。

（3）中央系统检验。

（4）巡视检验。

（5）抽取样本检验。

（6）指定样本检验。

（7）产前检验。

（8）全面的质量控制（品质保证）。

多数制衣厂都采用以上的一种或多种组合的方法。

第二节　质量标准

1　简介

在制衣业，质量指工作的标准与产出。拥有好品牌或知名商标的制造商，绝不会降低产品质量去追求价格、利润和产量，而是会寻求并使用更新、更好的制造方法，以便更有效地使用工厂的空间、人员和机器设备。

2 生产规格

在实际操作中，对于个别生产规格，应该在生产之前起草一个规格要点，涵盖缝制车间的所有作业环节，这将作为大纲并核对各独立工序所涉及的全部项目。下面的标题与例子便于车间编写缝制工序要点。

（1）部件定位：

①缝份。

②起止位置。

③回针。

④套结。

（2）搭配细节：

①对条对格。

②末端/边位/刀口/缝口的对位。

③带条/商标/衣架的位置。

④放松量/容位。

（3）针距细节：

①种类。

②针距密度。

③跳线/断线/漏针。

④张力外观。

⑤缝纫线颜色。

⑥线头长度。

⑦缝纫线损坏。

（4）完成后的外观：

①放松量。

②起皱。

③粘线。

④左右一致。

⑤条格。

⑥起褶。

⑦变形。

⑧原材料。

⑨正反面。

⑩色差。

⑪污渍。

⑫边/角位。

（5）熨烫工序：

①放松量。

②平服。

③起褶。

④变形。

⑤痕迹。

⑥黏合衬。

（6）标示/裁剪/刀口细节：

①形状。

②定位。

③模板。

（7）顾客需求：

①附加的特殊细节。

②外观的标准规格。

第三节　测量指南

1　上装

（1）后中长：放平服装并把后身对着自己。从后中领缝向下量至底边。

（2）肩宽：放平服装并把后身对着自己。确定肩缝与袖窿顶端点相交的位置，从肩点到肩点拉直量度。如果没有自然的肩缝，量度肩缝的自然折线与袖窿顶端点相交的位置。

（3）后背宽：放平服装并把后身对着自己。按照尺码表规定的测量点，从后中领位向下拉直量度两袖窿缝之间的长度。

（4）腰直：放平服装并把后身对着自己。从服装后中量度领缝到腰线的长度。

（5）袖长：放平服装，抹平袖子和肩部的皱褶，从后领中量起，经过肩点并沿着袖子中折线一直到袖口。

（6）袖克夫长：放平袖子（扣子款需扣好纽扣）。拉直袖口量度对折线之间的距离。

（7）胸围：放平服装，拉直服装袖窿底位，量度两端的距离。在服装胸围有松量时，在袖窿下侧缝处度量。

（8）下摆围：放平服装，并确保褶裥完全打开。在服装下摆处沿着下摆的形状量度。

（9）后领圈：从服装领圈自然状态下量度。

（10）袖窿：放平服装，对齐前后身袖窿缝并定位，以便使袖窿缝没有皱褶。沿着袖窿缝形状量度从袖窿顶端到底端的圆周尺寸。

（11）领长：敞开并放平服装领部，使服装内侧向着自己。沿着领座中线从纽扣中位量至扣眼的末端。

2　下装

（1）腰围：沿着腰线量度侧缝之间的距离。对于有裤腰的服装，沿着裤腰中位量度侧缝之间的距离。

（2）坐围：放平服装，根据尺码表的要求展开褶裥。抹平坐围处所有的皱褶。用软尺量度侧缝之间的距离。

（3）大腿围：放平裤子，展开大腿处的全部皱褶。用软尺从裤裆底横向直量至侧缝处。

（4）膝围：放平裤腿，并在膝围位平行裤子脚口量度折线之间的距离。

（5）裤脚宽：放平裤脚，垂直量度折线之间的距离。

（6）内长：放平裤子其中一条裤脚，抹平皱褶，并量度从裤裆点到裤子下摆的距离。

（7）外长：放平服装，并抹平所有皱褶。在同一侧缝量度从裤腰至下摆的距离。

（8）前裆：放平服装并将前身向着你自己，然后抹平上端部位的所有皱褶。沿着前裆量度从裤腰中点至裤裆的距离。

（9）后裆：放平服装并将后身向着自己，然后抹平上端部位的所有皱褶。沿着后裆量度从裤腰至裤裆的距离。

第四节　服装疵点的表述

1　牛仔裤疵点（表12-1）

表12-1　牛仔裤疵点

序号	描述	序号	描述
1	粗纱	16	后育克大小不均
2	染色不匀	17	缂袋缝头太大
3	部件色差	18	不正确的布层外露
4	裤腰跳线	19	后育克十字缝不对位
5	裤头高/低不均	20	袋布未伸至前门襟位
6	裤腰与前门襟不贴合	21	口袋缉线张力不均
7	裤身毛边	22	工字纽扣不牢
8	裤襻长短不均	23	前门襟宽度不均
9	裤襻遗漏	24	前门襟太长
10	套结缝在袋布上	25	裤腰扭曲
11	袋布太短	26	裤腰高度错误
12	内缝跳线	27	每英寸针迹比标准少
13	底层暴露	28	后袋形状不正确
14	口袋错位	29	下摆（脚口）扭曲
15	口袋边不直	30	里襟边位散口

续表

序号	描述	序号	描述
31	钻孔外露	36	洗水商标错误
32	左门襟缉线不均	37	裤裆底十字缝不对位
33	裤襻和样板不一样	38	拉链没延至顶端
34	套结遗漏	39	外缝缉线高低不平（落坑）
35	尺寸超出公差		

2 衬衫疵点（表12-2）

表12-2 衬衫疵点

序号	描述	序号	描述
1	领边不对称	23	口袋边位超出袋盖边位
2	领座不均	24	袖头缉线张力不均匀
3	领尖露齿（底）针迹	25	上领形状不均匀
4	下底层外露	26	机织商标偏离后中1/2英寸或更大
5	领缉面线张力不均	27	袖子褶裥距离不均匀
6	门襟起泡	28	绱袖头与袖子不贴合
7	领子由于熨烫不均匀而起泡	29	绱领子与衣身不贴合
8	门襟扭曲	30	下摆缝线滑移
9	袋子尺寸不正确	31	由于处理不正确导致领子起泡
10	领子缉面线后底层外露	32	下领嘴缝线断裂
11	纽扣疵点	33	表面打褶或起皱严重影响外观
12	领子尺寸不正确	34	纽扣缝制没有穿过所有孔眼
13	袖头边断裂	35	纽扣缝制时正面翻转
14	领尖不均匀	36	扣眼开口方向与规格相反
15	前门襟长度不均匀	37	1个或2个领插竹遗漏
16	绱袖起褶	38	缝型错误
17	熨烫欠佳	39	袋盖及袋盖高低不均
18	领尖缝线断裂	40	袖衩缝线断裂
19	止口错误	41	袖头下层冗余太多
20	袖头两边高度不均	42	折边不正确
21	缝纫线粘进领子内	43	袖头扭曲
22	扣眼线迹太窄		

第五节 检验报告

此表用于记录服装检查的资料。质量控制员或检验员检查货品时带上此表，找出存在的疵点并记录于表中。将此报告提交业务员、业务经理或厂长，以便做进一步改善提高

（表12-3～表12-7）。

表12-3　验货报告表

测量部位	规格	尺寸偏差S		规格	尺寸偏差M	规格	尺寸偏差L		内容	品质审核
										□中期　□尾期
前衣长									布料手感	
后中长									主标	
肩宽									尺码标	
胸围									洗涤商标	
腰围	73	+1	+2	77	√	83	-1		吊牌	
袖长									胶袋	
袖克夫宽									内盒	
袖窿围									外箱	
臀围	99	+1		104	√	109	-1	-2	整烫	
脚口宽	35	√		36	√	37	+1		搭配	
领围									外箱唛	
领高										
臂围										
前裆弧长（前浪）										
后裆弧长（后浪）										
大腿围										
裤长	100	+1		101	√	101	+1			

加工厂名：＿＿＿＿＿＿＿＿　　日期：＿＿＿＿＿＿＿＿

订单号：＿＿＿＿　货物名称：＿＿＿＿＿＿＿＿　货品总量：＿＿＿＿＿＿＿
货品号：＿＿＿＿　交货日期：＿＿＿＿＿＿＿＿　批号：＿＿＿＿＿＿＿

检验不符合规定之内容（疵点）：

工厂对不符合规定内容的改进对策：

工厂签名：＿＿＿＿＿＿　　验货员签名：＿＿＿＿＿

表12-4　宏达贸易有限公司

尺寸测量表（上装）												
卖方： _____　买主订单号： _____　合同号： _____　款式： _____												
批号： _____　数量： _____　检测者： _____　日期： _____												

测量部位	小码			中码			大码			加大码		
	规格	测量值	偏差	规格	测量值	偏差	规格	测量值	偏差	规格	测量值	偏差
前衣长												
后中长												
肩阔（宽）												
胸围												
腰围												
袖长												
袖克夫（介英）/ 袖口												
袖窿												
臂围												
下摆												
领围												
领高												
前领深												
后领深												
口袋												
帽子												

表12-5　宏达贸易有限公司

尺寸测量表（下装）												
卖方： _____　买方订单号： _____　合同号： _____　款式： _____												
批号： _____　数量： _____　检测者： _____　日期： _____												

测量部位	小码			中码			大码			加大码		
	规格	测量值	偏差	规格	测量值	偏差	规格	测量值	偏差	规格	测量值	偏差
腰围（放松量）												
腰围（拉伸量）												
上坐围												
下坐围												
大腿围												

续表

测量部位	小码			中码			大码			加大码		
	规格	测量值	偏差	规格	测量值	偏差	规格	测量值	偏差	规格	测量值	偏差
膝围												
裤口												
前裆弧长（前浪）												
后裆弧长（后浪）												
内缝长												
外缝长												
腰头高												
拉链长												

表12-6　品质控制部门

中期检查报告

卖方：_____　产品类型：_____　批号：_____　订单号：_____
装运：_____　数量：_____　日期：_____　检查人：_____

（Ⅰ）工艺方面：检查结果和生产改进的建议

检验不符合规定的内容（疵点）	数量	
	已检查数量	有疵点数量

（Ⅱ）合身程度：（a）不洗水/成衣普洗/石磨洗/服装染色　（b）洗水前/后　（c）熨烫前/后

测量部位	规格尺寸	规格尺寸	规格尺寸	规格尺寸	规格尺寸

（Ⅲ）生产状况：

裁剪：

缝制：

续表

后整理：	
包装：	
备注：	
（Ⅳ）卖方确认：_____	

表12-7　疵点清单——半成品检验报告

款号：_____				页号：_____	
面料疵点	严重疵点	轻微疵点	缝制疵点	严重疵点	轻微疵点
破洞			开缝		
污迹			缝口不牢		
疵点			毛边		
起球疵			起皱		
染色不匀			波形线迹		
烧斑			跳针		
横裆疵			断线		
			缝口错误		
成衣疵点	严重疵点	轻微疵点	成衣疵点	严重疵点	轻微疵点
布料颜色不匹配			底边不均		
部件颜色不匹配			鸳鸯片		
按扣有疵			部件缺失		
拉链有疵			格子不匀		
拉链带外露			纽扣松		
线头过多			扣眼不良		
熨烫疵点					

合格件数：_____	经检查的件数：_____
验收标准：_____	
拒收件数：_____	轻微疵件数：_____

日期	说明		质检员/跟单员所采取的行动	卖方签名
	生产状况			

日期	说明	质检员/跟单员所采取的行动	卖方签名

卖方有责任对检验中所发现的疵点进行更正并在此报告中做总结。然而，检验不解除卖方对货物装运过程中所发现的疵点应负的责任。

卖方签名：_____　日期：_____　检验员签名：_____　日期：_____

Garment Producing and Packaging
服装生产和包装

Chapter 13　Package and Sign in Garment
第十三章　服装包装和标识

项目名称： 服装包装和标识

项目内容： 1．服装包装功能

2．服装包装分类

3．服装标识

4．运输标志

5．装箱单

教学学时： 4课时

教学目的： 让学生了解服装包装的基本知识；了解服装包装的功能和分类；掌握服装外贸中包装标识。

教学方式： 通过课前任务，引入本章内容。讲解服装包装的功能和分类；通过拓展外贸案例，深入学习服装外贸中包装标识和运输标志；最后对本章学习难点和重点进行归纳，学生通过实操练习进一步熟悉和掌握本章节内容。

教学要求： （1）掌握本课词汇。

（2）了解服装中包装功能和分类。

（3）熟悉服装外贸中的包装标识。

（4）熟悉服装外贸中的运输标识。

Chapter 13 Package and Sign in Garment

Garment package refers to the general name of specific containers, materials and accessories used to protect the appearance and quality of garment products and facilitate the identification, sale and use of garment products in the process of transportation, storage and sales of garment products. In international trade, garment package is not only an important part of garment products, but also a main part of marketing strategy, which makes the products have competitive advantages. In recent years, it has also attracted extensive attention.

Part 1 Function of Garment Package

Garment package is the shell of garment goods, which is designed to protect all stages of the whole distribution. In addition to making products more attractive, package is an important tool for promotion.package in garment has the following basic functions.

(1) Protect garment, improve the beauty and grade of products, and attract consumers.

(2) Classify products of different batches, yards and models conveniently.

(3) Reduce storage space and save transportation cost.

(4) Introduce products, guide consumption and promote sales.

(5) Establish image and create value.

Part 2 Classification of Garment Package

According to the different functions of garment package in the circulation process, it can be divided into two types: sales package (i.e. inner package) and transportation package (i.e. outer package). The main function of the former is to protect goods and promote sales, while the latter is to protect goods and prevent goods damage and goods shortage.

Garment enterprises often attract customers through personalized brand package, such as paper bags, plastic bags, cartons and so on.

In the garment import and export trade, transportation package plays a very important role in transporting garment products from manufacturers to retail stores. The design of transportation package shall make it easy to transport, move and lift, and can stack regular shape package, such as standard specification wooden cases, cartons and vacuum package. Hanging or volumetric package is mainly used for the transportation and package of garment with high cost. In transportation package, various package methods such as barrels, bags and pallets will also be used.

Part 3   Signs of Garment

Signs of Garment is the communicator between buyers and products. Signs of Garment contains various types of information of the garment, such as buyer's name, country of origin, fabric composition, garment size, special care instructions, etc. garment without labels cannot be sold in foreign markets.

Signs of Garment is an important aspect of the production process. It is a part of the overall specification in the production process. The information contained in the logo is also important. garment identification generally includes brand label, content label, origin label, size label and washing label(Table 13-1).

(1) Brand label, also known as the main label, is the brand name or brand logo of the company that purchases and sells garment. Brand logo is a kind of "visual language". It transmits certain information to consumers through certain patterns and colors in order to identify brands and promote sales.

(2) Content label refers to the mark sewn on the garment to indicate the composition and content of fiber raw materials in the garment fabric and lining.

(3) Origin label refers to the name and address registered by textile and garment manufacturers according to law. Imported textiles and garment shall be marked in Chinese with the origin (country or region) of the product and the legally registered name and address of the agent or importer or seller in China.

(4) Size label refers to the model and size of garment.

(5) Washing label refers to helping customers understand how the product should be cared for. It indicates different types of care instructions such as washing, bleaching, drying, washing and ironing of garment. If targeted maintenance can be carried out, the garment will achieve higher durability and the color of garment will be more perfect within the time limit.

Table 13-1   Commonly Used Washing Marks

Label	Meaning
○	dryclean
⊗	not to dryclean
⊠	do not iron

Continued

Label	Meaning
	iron
	iron on low heat
	iron on medium heat
	iron on high heat
	line dry
	bleach
	do not bleach
	dry
	hang dry
	wash with cold water
	wash with warm water
	wash with hot water
	handwash only
	do not wash
	trouble dry with no heat
	do not trouble dry

Part 4 Transportation Labels

For most garment companies, the correct use of transportation labels has an amazing impact on the performance of package in the supply chain or delivery network. During transportation, the transportation labels on garment are very important,which are the instructions of the personnel transporting the goods. Transportation labels can be divided into shipping marks, indicative marks and warning marks.

(1)Shipping mark refers to the graphics, words and numbers written, stamped or brushed on the transportation package of goods. The main function is to facilitate the identification of relevant personnel in the process of loading, unloading, transportation and storage, so as to prevent wrong shipment and wrong transportation. Later, the simplified transportation mark formulated by the

United Nations Economic Commission for Europe only includes four contents: the prefix or abbreviation of the name of the consignee or buyer, the reference number (such as the sales contract), the destination (the name of the final destination or port of destination of the goods) and the number of pieces (the sequence number of each piece of goods in this batch and the total number of pieces of goods in this batch)(Fig. 13-1).

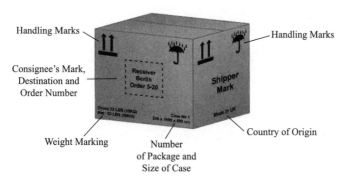

Fig. 13-1   Shipping Mark Sample

(2)Indicative marks are also called attention signs or safety signs. It is based on the characteristics of goods, in order to protect the goods and remind relevant personnel to pay attention in the process of transportation and storage. Indicative signs are generally marked on the package with simple and eye-catching graphics and words, such as "this end up", "handle with care", "keep dry", etc(Fig. 13-2).

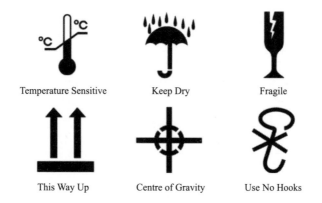

Fig. 13-2   Commonly Used Indicative Marks

(3)Warning marks: also known as danger warning sign, it is used to indicate that the goods are flammable, explosive, toxic, corrosive or radioactive and other dangerous goods. On the outer package, eye-catching graphics and words shall be used to warn relevant personnel to pay attention to prevent environmental pollution or personal injury. The mark must be painted on the obvious part of the package to make it clear at a glance(Fig. 13-3).

Fig. 13-3   Commonly Used Warning Marks

Part 5   Packing List

Packing List is made out by packing room. Information of the packing list should be the same as marked in the letter of credit(Table 13-2, Table 13-3). The information must contain:

(1)contract number.

(2)style number.

(3)description.

(4)production order.

(5)number of cartons.

(6)factory.

(7)delivery date.

(8)shipping mark, ect.

Table 13-2   Packing List

No.: _____

Contract No.: _____   P/O No.: _____

Style No.: _____   Description: _____

No. of Cartons: _____   Delivery Date: _____

From: _____   To: _____

Shipping Mark: _____

Unit: pc.

Item			Size and Quantities/Cartons	
C./No.	Color	Size of Carton	38 40 42 44 46 48 50	Total
1~15	Blue		12	× 15=180
16~25	Red		12	× 10=120

Table 13-3　Packing and Weight List

HONGDA GARMENT IMPORT & EXPORT CO., LTD				
PACKING AND WEIGHT LIST				
P/O NO. _____				

Order/Contract No.　　　　**Style No.**

No. of Cartons: _____

Quantities: _____

Marks and No.

Shipment Per. _____

Date: _____

Description:

Packing:

C./NO.	Ctns.	Total pcs.	Color				G.W.	N.W.	N.N.W.

SIZE OF CARTON:　　　　　G.W.

CU.FT.　　　　　N.W.

N.N.W.

New Words and Expressions

1. package   包裹；包
2. identification   识别；鉴定
3. transportation   运输
4. storage   储存；贮存
5. batch   一批；一组
6. consumption   消耗；消费
7. circulation   循环；流通
8. damage   损坏,破坏
9. classify   分类；划分
10. personalize   个人化
11. carton   纸箱；硬纸盒
12. retail   零售
13. stack   堆叠
14. standard   标准
15. specification   规格
16. wooden case   木箱
17. vacuum   真空
18. volumetric   立体的
19. barrel   桶
20. pallet   托盘
21. composition   成分；构成
22. mark   标识
23. country of origin   (原)产地
24. brand label   商标
25. size label   尺码标
26. washing label   洗水标
27. main label   主标
28. logo   商标
29. visual   视觉的
30. register   注册
31. bleaching   漂白
32. drying   干燥
33. ironing   熨烫
34. durability   耐久性；耐用性
35. handwash   手洗
36. indicative   直陈；指示的
37. warning   警告
38. flammable   易燃的
39. explosive   易爆炸的
40. toxic   有毒的
41. corrosive   腐蚀性
42. radioactive   放射性
43. eye-catching   引人注目的
44. environmental   环境的
45. pollution   污染
46. injury   伤害；损伤
47. L/C (letter of credit)   信用证
48. delivery company   货运公司
49. shipped   装运
50. P/O (production order)   生产制造单
51. carton   纸箱(外箱)
52. description   品名
53. C./No   纸箱编号
54. Ctns (cartons)   纸箱数
55. Pcs.(pieces)   件数
56. N.N.W (net. net weight)   净净重
57. CU.FT.(cubic foot)   立方英尺

Exercises

Translate the e-mail especially pay attention to the package expression in it.

DEAR NICO,

And now, I'm sending attached instructions for Shipping Marks and Packing Way—Brand H&G.

Important notes:

- Same as for BAS,you have to make inner pack by carton:1 STYLE / 1 COLOR / ASSORTED SIZES (FULL BRAKEDOWN).

- Each inner pack must be identified with a sticker showing STYLE + BARCODE.

- In this case is the barcode of the style by color, as in the Hang Tags.

-YOU CANNOT PACK ANY BROKEN (INCOMPLETE) PACK INSIDE THE CARTONS.

- All the boxes of same barcode should have same quantity.

So, I'm sending:

- Carton requested measures.

- Packing Way for all styles (Sheet 3&4: Pls note I left spaces for they to complete according to weight and measures instructed).Shipping Marks layout.

BEST REGARDS,

AMELIA

导入：服装外贸中包装功能是什么？

第十三章　服装包装和标识

服装包装是指在服装产品运输、储存、销售的过程中，用于保护服装产品外观和质量，以及为了便于识别、销售和使用服装产品而使用的特定容器、材料及辅助物等物品的总称。国际贸易中，服装包装不仅是服装产品的重要组成部分，也是营销策略的一个主要组成部分，使产品具有竞争优势，近年来也受到了广泛关注。

第一节　服装包装功能

服装包装是服装商品的外壳，它的设计目的旨在保护整个销售的各个阶段。包装除了使产品更具有吸引力，更是一种重要的促销工具。服装包装具有以下基本功能。

（1）保护服装、提升产品的美观程度和档次，吸引消费者。

（2）便于对不同批次、码数、型号的产品进行分类。

（3）减少储存空间，节省运输成本。

（4）介绍产品、指导消费、宣传促销。

（5）树立形象、创造价值。

第二节　服装包装分类

根据服装包装在流通过程中所起作用的不同，可分为销售包装（即内包装）和运输包装（即外包装）两种类型。前者的主要作用在于保护商品和促销，后者主要在于保护商品和防止出现货损货差。

服装企业经常通过个性化的品牌包装来吸引顾客，例如纸袋、塑料袋、纸盒等。

在服装进出口贸易中，要想把服装产品从生产厂家运到零售店，运输包装起着非常重要的作用。运输包装的设计应使其易于运输、移动和搬起，可以堆叠规则形状的包装，例如标准规格的木箱、纸箱、真空包装。挂装或立体包装主要是针对成本较高的服装所采用的运输包装方式。在运输包装中，还会用到桶类、袋类、托盘等包装方式。

第三节 服装标识

服装标识是买家和产品之间的沟通者。服装标识包含该服装的各种类型的信息，如客户名称、原产国、面料成分、服装尺码、特殊护理说明等。没有标识的服装不能在国外市场销售。

服装标识是生产过程中的一个重要环节。它是生产过程中整体规范的一个部分。标识包含的信息也很重要。服装标识一般包括品牌标识、成分标识、产地标识、尺码标识和洗涤标识（表13-1）。

（1）品牌标志：也叫作主标，是表示采购和销售服装的公司的品牌名称或品牌标志。品牌标志是一种"视觉语言"。它通过一定的图案、颜色来向消费者传输某种信息，以达到识别品牌、促进销售的目的。

（2）成分标识：是指缝制在服装上表明服装面料和里料中纤维原料成分及其含量的标志。

（3）产地标识：是指纺织品和服装生产企业依法登记注册的名称和地址。进口纺织品和服装应用中文标明该产品的原产地（国家或地区）以及代理商、进口商或销售商在中国依法登记注册的名称和地址。

（4）尺码标识：是指表示服装的型号、尺码等。

（5）洗涤标识：是指帮助客户了解产品应该如何护理。标示衣物的洗涤、漂白、烘干、洗涤、熨烫等不同类型的护理说明。如果能够有针对性地进行保养，则服装耐用性更高，服装色泽更完美。

表13-1 常用洗涤标识

标识	含义
◯	干洗
⊗	不可干洗
⊠	不可熨烫
⊟	熨烫
⊟	低温熨烫（100℃）
⊟	中温熨烫（150℃）
⊟	高温熨烫（200℃）
⊍	洗涤
△	可漂白

续表

标识	含义
▲	不可漂白
▢	悬挂晾干/干衣
⊞	随洗随干
⊻	冷水机洗
⊻	温水机洗
⊻	热水机洗
⊻	只能手洗
⊠	不可洗涤
◉	无温转笼干燥
⊠	不可转笼干燥

第四节　运输标志

对于大多数服装公司来说，正确使用运输标志对包装在供应链或交付网络中的表现有着惊人的影响。运输时，服装上的运输标志非常重要，这些标志是运输货物人员的指示。运输标志可分为装运标签、指示标志和警示标志。

（1）装运标签：又称作唛头，是指在商品的运输包装上书写、压印或刷制的图形、文字和数字。主要作用是便于装卸、运输、储存过程中有关人员的识别，以防止错发错运。后来，联合国欧洲经济委员会制定的简化运输标志中只包括四项内容：收货人或买方的名称字首或缩写、参考编号（如买卖合同）、目的地（货物运送的最终目的地或目的港的名称）和件数号码（本批每件货物的顺序号和该批货物的总件数）（图13-1）。

图13-1　唛头样例

（2）指示标志：又称注意标志或安全标志。它是根据商品的特性，为了保护好商品，提醒有关人员在运输和储存过程中加以注意的事项。指示性标志一般都是以简单醒目的图形和文字在包装上标明，如"此端向上""小心轻放""保持干燥"等（图13-2）。

图13-2　常用指示标志

（3）警告标志：又称危险警告标志，用于表明货物是易燃、易爆、有毒、有腐蚀性或有放射性等危险性的物品。在外包装上以醒目的图形和文字警示有关人员加以注意，防止造成环境污染或人身伤害。标志必须印刷在包装上的明显部位，使人一目了然（图13-3）。

图13-3　常用警告标志

第五节　装箱单

装箱单由包装车间填写。装箱单内容必须与信用证上标注的一致（表13-2、表13-3）。内容必须包括：

（1）合约编号。

（2）款号。

（3）品名。

（4）生产通知单。

（5）装箱数。

（6）制造厂。

（7）交货期。

（8）箱唛等。

表13-2　装箱单

编号：_____	
合约编号：_____	制单号：_____
款号：_____	品名：_____
箱数：_____	交货日期：_____
从_____	至_____
箱唛：_____

单位：件

项目			尺码、数量/箱	
箱号	颜色	纸箱尺寸	38、40、42、44、46、48、50	总计
1-15	蓝色		12	×15=180
16-25	红色		12	×10=120

表13-3　包装与重量清单

宏达服装进出口有限公司

包装与重量清单

制单号：_____

订单号/合同号：	款号：	箱数：
		数量：

箱唛和编号：

运输方式 _____
_____
_____

日期：_____

品名：

装箱要求：

续表

箱号	箱数	总件数	颜色				毛重	净重	净净重

纸箱尺寸:	毛重:
立方英尺:	净重:
	净净重:

本模块微课资源（扫描二维码观看视频）

9.1 Seam Allowance & Front Edge & Ease & Fullness的区分

9.2 What is King Size?

9.3 Abbreviation for Garment

10.1 Equipment for Garment

11.1 Welt Pocket

11.2 Fly in Garment

12.1 English Expression of Garment Defects

12.2 English Expression of Fabric Patterns

13.1 Washing Signs in Garment

13.2 Package and Sign in Garment

13.3 Transportation Signs

Module 4
Communication in Garment Foreign Trade

模块四
服装外贸交流

Communication in Garment Foreign Trade
服装外贸交流

Chapter 14　　Business E-mails
第十四章　　商业电子邮件

项目名称：商业电子邮件

项目内容：1. 商业电子邮件简介

　　　　　2. 商业电子邮件格式

　　　　　3. 服装外贸商业电邮案例

教学学时：3课时

教学目的：让学生了解服装外贸邮件的格式，具备函电的阅读、翻译和书写能力。

教学方式：由教师通过课前导入任务引出本章内容。了解外贸邮件，分析讲解服装外贸邮件的格式并总结归纳学习难点和重点，结合服装外贸商业电邮案例进行深入学习。

教学要求：（1）熟悉外贸邮件的格式和撰写技巧。

　　　　　（2）熟悉阅读、翻译和撰写服装外贸商业电邮。

Chapter 14   Business E-mails

Part 1   Introduction to Business E-mails

E-mail is the abbreviation of electronic mail, and it is a method of sending messages to one or more people through a computer network. In garment enterprises, many branch offices or departments of large companies will have already been familiar with sending electronic messages via their company computer network.

The e-mail needs to contain the following information:

(1) To: The e-mail address that you want to send the message to.

(2) Subject: A brief and meaningful statement about the contents of the message allows the receiver to read quickly through his list of incoming mail and identify messages of special importance.

(3) Cc: Addresses of those people who will receive a copy of the message. "Cc" refers to "Carbon copy".

(4) Body Text: Start this part with a salutation, and follow by the message. And then finish with a complimentary close and signature.

(5) Attached File: Most e-mail allows you to attach word processed documents, program files or pictures, etc. Whenever you send an attachment, it is advisable to include a short introductory sentence explaining the contents of the attached file.

Part 2   The Format of E-mail Analyzing

Totally, we have to respect the following guidelines when composing e-mail messages:

(1) Keep one subject in the same message.

(2) Use a descriptive and informative subject heading.

(3) Keep messages short and indicate with point by point.

(4) Write short sentences.

(5) Quote from the original e-mail if required and target dates.

Date: May 19, 2020

To: ivanzxl@126.com

Cc: ray@tom.com

From: may166@gmail.com

Subject: Details about carton box for style M-12

Dear ivanzxl,

　　Please follow up the details about carton size of style M-12.

　　The difference in the hanger should not affect the height of the carton, so if we change to the A4 carton (21" *15" *15"), which is about an inch higher, can we get at least one more garment in a carton?

　　If we make this change, I would also like to change the style M-13 to the same A4 carton. How many units of a M-13 can you get into an A4 carton?

　　Please confirm your acceptance by return.

　　Thanks.

Yours faithfully,

May

Merchandiser

Polo Trading Company

Date: Friday, December 28, 2020, 20:11:48

To: Ray@pub.huizhou.gd.cn

Cc: Bandy168@163.net

From: rykiel@winpap.co.uk

Subject: Special Pre-pricing

Dear Sir,

　　Because the rising in the world price of cotton, from January 1 of next year, prices for our cotton lines are due to increase by 10% across the board.

　　As you are a valued customer of long standing, we wish to give you the opportunity to impact the price increases by ordering now at the current prices.

　　In addition, we are willing to give you a discount of 10% on all orders of more than HK$6,000.

　　We believe that you will see the advantage of this arrangement, which will save you at least 10% on cotton fabric purchases in the coming season.

We look forward to your early reply.

Yours faithfully,

Alice Wong

Export Sales Manager

Rykiel Trading Company

Part 3　E-mail Cases in Garment Foreign Trade

Commercial e-mail in garment foreign trade mainly includes: inquiry e-mail, offer e-mail, counter-offer e-mail, order confirmation e-mail and agreement signing e-mail.

1　E-mail about enquiry

Have the simplicity & clearing as a rule in the general enquiry letter. The content must be shown category list, price list, unit price and sample swatch, etc.

Dear Sir,

We should be glad to know whether you supply lots of 100% cotton men's shirts, blouses, shorts, and trousers, etc.. We also require lots of bath towels, please send us the category of some goods that you recommend or supply from stock.

Yours faithfully,

Alice Wong

Export Sales Manager

Rykiel Trading Company

2　E-mail about offering goods

The offering e-mail or sales email is very difficult to write. A nice sales e-mail has four characteristics: arouse interest, create desire, carry conviction and induce action. And it is most important to show the necessary information with offering goods.

(1) Style.

(2) Order date & Order No..

(3) Category No..

(4) Size & Color.

(5) Quantity.

(6) Unit price & Discount.

(7) Total price.

(8) Payment.

(9) Delivery.

(10) Signature.

Dear Madam,

　　We are pleased to receive your inquiry about price about reversible wool blankets, 15% wool blended with 85% cotton. We will offer you as the following:

Name of commodity	Reversible wool blankets, 15% wool blended with 85% cotton
Unit price	HK$140 on FOB basis in Chicago
Quantity	200,000 pieces
Color	White and black
Terms of payment	30%by T/T, the balance by T/T within one week after receiving the copy of B/L
Delivery date	Within 15-25 days after signing the contract

　Look forward to hearing from you.

Yours faithfully,

Alice Wong

Export Sales Manager

Rykiel Trading Company

3   E-mail about counter offers

A reply to an offer which purports to be an acceptable but contains additions, limitations or other modifications is a rejection of the offer and constitutes a counter-offer. The counter offer should basically include the contents of following four parts:

(1) Express thanks for the offer (indicating the date and the content of the offer).

(2) Express regret of inability to accept the offer, stating the specific terms and giving the reasons.

(3) Put forward your own terms and conditions.

(4) Express the wish that the correspondent can accept the counter offer.

Dear sir or madam,

　　Thank you for your offer of April 21, offering us for reversible wool blankets, 15% wool blended with 85% cotton HK$140 on FOB basis in Chicago.

　　But we are regret to tell you that we can't accept your price at HK$140 on FOB basis.

At our end, we wish the business can be done at HK$110 on FOB basis. We require payment by D/P at sight. We wish your counter-offer is acceptable to you, and we are looking forward to hearing from your soon.

Yours faithfully,
Alice Wong
Export Sales Manager
Rykiel Trading Company

4 E-mail about order confirmation

When confirming an order received, the following structure may be for your reference:

(1)Express appreciation for the order received.

(2)Assure the buyers that the goods they have ordered will be delivered in compliance with their request. It is also advisable for the sellers to take the opportunity to resell their products or to introduce their products to the buyers.

(3)Close the letter by expressing willingness to cooperate or suggesting future business dealings.

Dear sir or madam,

Thank you so much to receive your confirmed order. I assure the goods that you have ordered will be delivered in compliance with your request. If it is possible, you can also see our other goods to see whether you need it or not. I wish that we will keep cooperating and business dealings next time.

Yours faithfully,
Alice Wong
Export Sales Manager
Rykiel Trading Company

5 E-mail about signing contract

Dear sir or madam,

Thank you for your e-mail and we appreciate it that you accept our counter offer. We understand you standing and agree to pay by L/C. We hope we can do T/T in the future. Enclosed pls. find our order. I will send you the original one today of which pls. countersign one copy for our file. We need four approval samples before your mass production. When shall I receive them? Any questions, just feel free to ask me.

Yours faithfully,

Alice Wong

Export Sales Manager

Rykiel Trading Company

New Words and Expressions

1. message   消息；信息
2. statement   陈述;说法
3. salutation   称呼
4. complimentary close   结尾敬语
5. signature   署名；签名
6. body   正文
7. attach   粘贴；附上
8. attachment   附件
9. subject   主题
10. pre-pricing   预定价
11. abbreviation   缩写,略语
12. yours faithfully   敬上(用于商业书信)
13. unit price   单价
14. sample   样板
15. commodity   商品名称
16. payment   付款方式
17. reversible   正反面
18. attach file   附件
19. merchandiser   跟单员
20. look forward to   期望
21. discount   折扣
22. blanket   毛毯
23. inquiry   询盘
24. offer   报盘
25. counter-offer   还盘
26. confirmation   确认
27. simplicity   简明
28. category list   目录表
29. price list   价钱单表
30. unit price   单价
31. swatch   (织物的小块)样品；布样
32. addition   补充；添加
33. limitation   限制；局限
34. modification   更改；修改
35. rejection   拒收
36. constitute   构成；组成
37. inability   不能
38. put forward   提出
39. correspondent   相应的
40. appreciation   感激
41. willingness   意愿
42. countersign   副署；会签

Exercises

Write a piece of e-mail about offer goods in the garment industry by your suggestion.

导入：如何写一封服装外贸的询价电子邮件？

第十四章　商业电子邮件

第一节　商业电子邮件简介

E-mail是电子邮件的缩写，且它是通过计算机网络将信息发给一个或更多人的一种方法。在服装界，很多大公司的分公司或部门已经熟悉通过公司的计算机网络传递电子邮件。

电子邮件需包括如下资料：

（1）收件人地址：你需要发信息过去的地址。

（2）主题：在发件人快速浏览所收到的信件时，一个简明扼要的标题会让他立刻注意到关键信息。

（3）抄送人地址：那些同样能将收到信息副本的地址。"Cc"是指"副本"。

（4）信文：这部分必须以称呼作为开始，接着是信息内容。然后以结束语与签名结束。

（5）附件：多数电子邮件允许你附上文字文件、程序文件、图片等。当发送附件时，最好包含一份简短介绍用来解释附件的内容。

第二节　商业电子邮件格式

当我们编写电子邮件信息时，必须注意以下几点：

（1）同一邮件保持一个主题。

（2）采用描述性与资料性的主题标题。

（3）信息保持简短，并以要点形式指出。

（4）书写简短的句子。

（5）如果需要和预定日期，要引用原件发送。

日期：May 19，2020

收件人：ivanzxl@126.com

副本转呈：ray@tom.com

发件人：may166@gmail.com

主题：有关M-12款的纸箱细节

亲爱的ivanzxl：

请跟进M-12款纸箱的有关尺码细节。

衣架的不同不会影响纸箱的高度，如果我们改为用A4纸箱（21英寸×15英寸×15英寸），纸箱大约高了1英寸，能否至少多放一件服装？

如果我们做出改变，我们也想将M-13款改为同样的A4纸箱。一个A4纸箱能放多少件M-13款的服装？

请回信确认你们的接受情况。

谢谢！

May

跟单员

Polo 贸易有限公司

日期：Friday，December 28，2020，20：11：48

收件人：Ray@pub.huizhou.gd.cn

副本转呈：Bandy168@163.net

发件人：rykiel@winpap.co.uk

主题：特殊的预定价

亲爱的先生：

由于全世界的棉产品价格一直在上涨，从明年1月1日起，我们的棉产品系列价格将全面上涨10%。

因您是长期支持我们的忠诚顾客，在价格上涨前我们给贵公司提供一个机会，以现在的价格订购货品。

另外，如果您的订单总额超过6000港元，我们将给予10%的折扣。

我们相信您能看到这项安排的优点，提早订货，与下个季节购买全棉布料相比至少可减少10%的支出。

我们希望您能尽早答复！

敬上

Alice Wong

出口部经理

Rykiel贸易公司

第三节　服装外贸商业电邮案例

服装外贸商业电邮主要包括：询盘邮件、报盘邮件、还盘邮件、订单确认邮件和签订协议邮件。

1.询盘邮件

普通的询盘邮件应以简洁和清晰为原则。内容必须显示目录表、价目表、单价与样板等。

亲爱的夫人：

　　我们很高兴收到你的询问信件，询问有关15%羊毛混纺和85%棉的双面羊毛毯子价格。我们能给你的单价为FOB价140港元，在芝加哥交货。请给我们寄一些你们推荐的产品目录或仓库的现货。

　　敬上

Alice Wong
出口部经理
Rykiel 贸易公司

2.报盘邮件

提供商品信件或销售信件非常难写，一份好的销售信件有四个特征：引起兴趣、创造需求、令人信服和引导行动，且最重要的是要显示出所提供商品的必要信息。

（1）款式。

（2）订单日期与订单编号。

（3）目录编号。

（4）尺码与颜色。

（5）数量。

（6）单价与折扣。

（7）总价。

（8）付款方式。

（9）交货期。

（10）签名。

亲爱的夫人：

我们很高兴收到你的询问信件，询问有关15%羊毛混纺和85%棉的双面羊毛毯子价格。我们现在能给你的报盘如下：

商品名称	15%羊毛混纺85%棉的双面羊毛毯子
单价	FOB价140港元，在芝加哥交货
数量	200，000张
颜色	白色和黑色
付款方式	以电汇先付3成，余款收提单后十个工作日内付请
交货日期	签订合同后15～25个工作日内发货

我们希望您能尽早答复！

敬上

Alice Wong

出口部经理

Rykiel 贸易公司

3.还盘邮件

对发盘表示接受但有添加、限制或其他更改的答复，即为拒绝该项发盘，并构成还盘。还盘应包含以下四个方面的内容：

（1）对发盘报价表示感谢（注明报价日期和内容）。

（2）说明无法接受的条件，以及原因或者要求变更装运期或支付条件。

（3）提出我们的条款和条件。

（4）表达成交的愿望。

亲爱的先生或女士：

非常感谢收到你关于羊毛的、15%羊毛混纺85%棉的毯子FOB 价140港币在芝加哥交货的发盘。但很遗憾我们不能接受你提出的价格，我希望我们可以以FOB 价110港币在芝加哥交货成交。

希望你方能够接受我们的还盘，期待尽快收到您的回复。

敬上

Alice Wong

出口部经理

Rykiel 贸易公司

4.订单确认邮件

当我们收到关于订单确认的电邮时，电邮的内容应该包含以下三个部分：

（1）表达收到订单的感激。

（2）向买方保证，买方所订购的货物一定会按照买方要求按时交付，有机会可以向买方推荐一下卖方自己的其他货物。

（3）最后表达未来继续合作的意向。

亲爱的先生或女士：

非常感谢收到你的确认订单，我保证你所订购的货物一定会按你方的要求按时交货，你也可以看一下我们其他的货物，看是否需要。非常期待我们下次能继续合作。

敬上

Alice Wong

出口部经理

Rykiel 贸易公司

5.签订协议邮件

亲爱的先生或女士：

非常感谢你接受我们的还盘。关于支付条件，我们理解你的处境，我们接受信用证付款的条件。希望以后我们做T/T。附件是我们的订单，请查收。正本将于今天寄给你，你收到后请签署寄回一份给我。我们要求你在生产前，寄四个样板给我们确认后方可生产，请问样板何时可以寄给我们确认。如有问题，请与我联系。

敬上

Alice Wong

出口部经理

Rykiel贸易公司

Communication in Garment Foreign Trade
服装外贸交流

Chapter 15   Practical Dialogue for Foreign Trade Garment
第十五章   服装外贸中的沟通交流

项目名称： 服装外贸中的沟通交流

项目内容： 1. 服装产品的沟通交流

2. 样衣的沟通交流

3. 谈判和签订合同的沟通交流

4. 质量和投诉的沟通交流

教学学时： 3课时

教学目的： 让学生掌握在各种服装外贸情境下的对话，培养学生的服装外贸口语表达和会话能力。

教学方式： 由教师通过课前导入任务引出本章内容。分析讲解服装外贸情景下的对话；结合服装外贸案例，针对服装外贸情境，分组开展对话演练；归纳本章学习难点和重点。

教学要求： （1）掌握在各种服装外贸情境下的对话。

（2）培养学生的服装外贸口语表达和会话能力。

Chapter 15　Practical Dialogue for Foreign Trade Garment

Part 1　Communication on Garment Production

Dialogue 1

(M—Merchandiser; C—Customer)

M: Hello. What can I do for you?

C: Do you have new silk samples?

M: Yes, this is new sample.The style is very popular among foreign customers.

C: How many colors?

M: Navy blue, black, blue, and coffee.

C: Okay, I think the shell fabric needs further improvement of comfort, drape, fullness and color.

M:You get what you pay. Compared with this price, the quality of this garment is quite good.

C: I'll think it over.

Dialogue 2

(M—Merchandiser; C—Customer)

M: How about this overall jeans?

C: The design and the color are quite OK, but I don't like the material.

M: You can change other material.

C: Do you have chino material?

M: Yes, I show you.

C: Style and material are ok. What colors do you have?

M: Caramel, violet and lemon yellow. Three colors. You can order other colors.

C: What colors sell well?

M: Bright colors are trend color all over the world this year.

C: Ok. Violet and lemon yellow.

M: What colors do you want?

C: Each color 500 pieces.

M: No, each color at least 1, 000 pieces.

C: Much more than what I was expect.

M: One step back, at least 800 pieces.

C: Deal.

Part 2   Communication on Garment Samples

Dialogue 1

(M—Merchandiser; C—Customer)

M: This is the first sample we submitted.The specifications of sample size and raw material are listed on sample order.

C: It seems that the flap is too small and a little too narrow across the shoulder, please amend the size.

M: Sorry, this is our mistake.

C: The thread should be 12−14 needles per 3 cm, the sample is only 10 needles per 3 cm.

M: We will inform the Sample Room to revise these points.

C: Okay. Please use CVC 60/40 260g/M2 brushed fleece to make approving sample.

M: We' ll send approving samples in each color with each size before November 1st.

C: That' s great. Thank you.

M: You are welcome.

Dialogue 2

(M—Merchandiser; C—Customer)

C: Here, the hook & bar of sample jeans are different from the origin sample. Pls check.

M: Sorry. We' ll do that asap.

C: And these zippers here need replacement. Must use our appointed supplier.

M: Uh−huh. What else?

C: For the shipping sample problem, pls. advise Jimmy. He is in charge of it.

M: Of Course. I' ve written it down.

Part 3   Communication on Garment Negotiation and Contract

Dialogue 1

(M—Merchandiser; C—Customer)

C: Can I order it?

M: Yes, but it will take about two weeks.

C: Oh, I can' t wait that long. Why so long?

M: We are in the busy period. I am sorry, but there is nothing we can do about it. But if you can pay a deposit first, We' ll reserve it for you.

C: OK. How to .pay a deposit?

M: We ask for 30% down payment. And you will pay the balance on delivery.

C: No problem.

Dialogue 2

(M—Merchandiser; C—Customer)

C: Are there any stock left?

M: Yes, there are stock.

C: How many pieces are in the stock?

M: 5000 pieces.

C: How much?

M: 25 Yuan.

C: No, I will take everything in stock. Give me the last price.

M: Well, 23 yuan.

C: The price is too high. How about 20 yuan?

M: Sorry, that' s almost cost price. Let' s meet half way. 22 yuan, OK?

C: All right. I' ll take it.

M: Thanks.

Part 4　Communication on Garment Quality and Complaints

Dialogue 1

(M—Merchandiser; C—Customer)

M: Good morning, sir. Can I help you?

C: I hope so. I' d like to complaint. The goods you sent to us seem to be out of joint with your samples.We can' t accept the articles as they are not equal to samples.

M: what' s the matter?

C: It' s not what I ask for.

M: What' s wrong with it?

C: You are not up to the agreed specifications and quality.

M: Really?

C: The pattern is not right, either.

M: Let me check out...The leather is of the same quality, but this is hand-made, so the design and color of each piece will be different.

C: Only first-class goods will be accepted. I'm afraid we can't accept it.

M: I quite understand you, and we will follow up later.

C: All right.

Dialogue 2

(M—Merchandiser; C—Customer)

C: I am afraid I'll have to cancel our order.

M: What's the matter?

C: The shipment is not consist with the original sample.

M: These garments were produced after the fashion of the sample given by your company.

C: We regret to say that we are not in a position to accept these garments as they are out of accord with the sample.

M: You can't be serious.

C: No, I am serious. I am afraid I'll have to cancel the deal unless you reduce your price.

M: Well, in view of our good cooperation over the past years, we can take some considerations, but only for this order.

C: Then how much can you go down?

M:0.5% off the original price.

C: 0.5%! your reduction is too modest. What about 2%?

M: No. I am afraid it's still not acceptable.

C: OK. What do you say to 1%, then?

M: I have no choice but to accept your condition.

C: Thank you. Let's call it a deal.

M: All right. That's settled.

New Words and Expressions

1. comfort 舒适性
2. drape 垂悬性
3. fullness 丰满度
4. overall jeans 背带牛仔裤
5. trend 流行；趋势
6. deal 成交
7. submit 提交
8. raw material 原材料
9. brushed fleece 抓毛绒布
10. replacement 替代；替换
11. deposit 定金
12. reserve 保留；预留
13. balance 余额；余款
14. down payment 头期款

15. stock　库存
16. last price　最低价
17. cost price　成本价
18. appoint　指定；任命

19. supplier　供应商；供货商
20. hand-made　手工的
21. first-class goods　一等品

Exercises

Make a dialogue with your partners on garment quality and complaints in foreign trade.

导入：如何和客户交流服装的材料、图案、款式和色彩？

第十五章 服装外贸中的沟通交流

第一节 服装产品的沟通交流

对话一

（M：跟单员；C：顾客）

M：你好。我能帮你什么？

C：你们有新款的真丝样衣？

M：这就是新款。在国外客户中非常受欢迎的一款。

C：几种颜色？

M：藏青色、黑色、蓝色和咖啡色。

C：我认为面料的舒适性、垂悬性、丰满度及色泽等有待进一步提高和改进。

M：俗话说，一分钱一分货。和这个价钱比起来，这款衣服的质量算是不错的了。

C：我考虑一下。

对话二

（M：跟单员；C：顾客）

M：这件背带牛仔裤怎么样？

C：款式和颜色都不错，但我不喜欢这种面料。

M：你可以换其他面料。

C：你们有斜纹面料吗？

M：有的，我拿给你看看。

C：款式和面料都可以。有什么颜色？

M：驼色、紫色和柠檬黄，三种颜色。你可以定这些色。

C：什么颜色卖得好。

M：今年全球流行亮色。

C：紫色和柠檬黄。

M：每色要多少件？

C：每色500件。

M：每色至少要1000件。

C：这超出了我的预期。

M：一人退一步，至少要800件。

C：成交。

第二节　样衣的沟通交流

对话一

（M：跟单员；C：顾客）

M：这是我们提交的头板。样衣尺码和原材料都列在样板单上。

C：看起来袋盖太小，肩宽有点窄，请修改尺寸。

M：抱歉，是我们的失误。

C：线迹应该是每3厘米12～14针，样衣只有每3厘米10针。

M：我们通知样品室修改。

C：好。请用CVC60/40 260g/M2拉绒面料做确认样品。

M：我们将在11月1日前寄出每种颜色、每种尺寸的合格样品。

C：太好了，谢谢。

M：不客气。

对话二

（M：跟单员；C：顾客）

C：这里，样品牛仔裤裤钩和原样品不同。请核对。

M：抱歉，我们尽快核对。

C：还有这些拉链要换。必须用我们指定的供应商。

M：还有其他的吗？

C：关于船样的问题，请联系吉米。他负责这件事。

M：好的，我都记下来了。

第三节　谈判和签订合同的沟通交流

对话一

（M：售货员；C：顾客）

C：我能订购吗？

M：当然可以，不过大概需要两个星期的时间。

C：喔，不行，我不能等那么久。为什么要那么久呢？

M：我们这段时间是旺季。很抱歉，实在是没办法。如果你交预付款，我们可以帮你留货。

C：好吧，预付款怎么支付？

M：我们要求30%的定金。在交货时付清余款。

C：没问题。

对话二

（M：售货员；C：顾客）

C：还有库存吗？

M：是的，还有库存。

C：还有多少件库存？

M：5000件。

C：多少钱？

M：25元。

C：不，我要全部库存，给我最低价。

M：好的，23元。

C：价格太高了，20元怎么样？

M：抱歉，这是成本价了。我们各退一步，22元，好吧？

C：好吧，我要了。

M：多谢。

第四节　质量和投诉的沟通交流

对话一

（M：售货员；C：顾客）

M：早上好，先生。有什么能帮您吗？

C：但愿能帮上，我想投诉。你们发来的货物似乎与你们的样品不符合。我们不能接受
这些货物，因为它们与样品不符。

M：怎么回事？

C：这不是我想要的。

M：哪里不对？

C：你方达不到商定的规格和质量。

M：真的吗？

C：样式也不对。

M：让我查一查。皮的质量是一样的，但这件是手工制作的，所以每一件的花色会
不同。

C：只有一等品才能被接受。恐怕我们无法接受。

M：我很理解你，我们稍后会跟进。

C：好的。

对话二

（M：跟单员；C：顾客）

C：恐怕我必须取消我们的订单。

M：怎么回事？

C：这批货物与原样品不一致。

M：这些服装是按照你们公司提供的样品生产的。

C：很遗憾，我们不能接受这些服装，因为它们与样品不符。

M：您不是在开玩笑吧。

C：不，我是认真的。恐怕我不得不取消这笔交易，除非你降价。

M：好吧，鉴于过去多年的友好合作，我们可以做些考虑，但仅限这批订单。

C：那么你能减多少？

M：原价的0.5%。

C：0.5%！你的降价幅度太小了。2%怎么样？

M：不行。恐怕我还是不能接受。

C：好吧，那您觉得1%怎么样？

M：除了接受你的条件，我别无选择了。

C：谢谢。那么成交了

M：好的，就这么定了。

Communication in Garment Foreign Trade
服装外贸交流

Chapter 16　Garment E-commerce
第十六章　服装电子商务

项目名称：服装电子商务

项目内容：1．服装电子商务

　　　　　2．服装电子商务平台

　　　　　3．如何提升外贸业绩

教学学时：3课时

教学目的：让学生了解电子商务、各电商平台的优缺点，学会利用电商平台提升服装外贸业绩。

教学方式：由教师通过课前导入任务引出本章内容。服装电子商务的概念、特点和优缺点。结合案例，针对服装外贸电商化、网络化趋势进行讨论。

教学要求：（1）掌握本课词汇。

　　　　　（2）熟悉服装电子商务平台。

　　　　　（3）熟悉用电商平台提升服装外贸业绩技巧。

Chapter 16　Garment E-commerce

Part 1　Garment E-commerce

Garment industry and E-commerce, a traditional industry and a new industry, have some certain natural connections after several years development.

E-commerce is from electronic commerce, abbreviated as EC. As one of the new marketing means of garment enterprise, E-commerce is more and more popular because of its convenience, low-cost and rapid spread with the development of information and internet.

In China, garment e-commerce has experienced three major development opportunities: the first was SARS in 2003; the second was the financial crisis in 2008; the third is the COVID-19 epidemic in 2020.The traditional garment industry has been developing blowout in the new network sales channel, and the garment business is in its mature stage.

Part 2　Garment E-commerce Platform

Most people know shopping online from B2B firstly ,then promoting the B2C and C2C development.The new B2C marketing model, the connection of traditional industry and E-commerce, is started by PPG, such as Vancel, Mecox Lane etc. Nowadays, garment e-commerce platforms show the following trends: the boundary between B and C is becoming more and more blurred, with the emergence of B2B2C, C2C2B and so on; C2B this customized form, with the personalized consumption, the demand is increasing(Table 16-1).

Table 16-1　Garment E-Commerce Platforms

No.	Abbr.	Full Name	Platform
1	B2C	Business to Customer	JD.com, tmall.com, vipshop, Taobao Mall, Amazon, VANCL
2	B2B	Business to Business	1688. Alibaba international station
3	C2C	Consumer to Consumer	Taobao, paipai

For strong foreign trade enterprises, through the form of vertical e-commerce and cross-border e-commerce, they can get rid of the constraints of large-scale third-party cross-border e-commerce platforms, and build and operate their own cross-border e-commerce platforms.According to the characteristics and development needs of their own foreign trade business, they have the following advantages: reducing costs, business specialization, meeting personalized needs, etc(Table 16-2).

Table 16-2　Platform E-Commerce, Vertical E-Commerce and Cross-Border E-Commerce

No.	Name	Characteristic	Platform
1	Platform E-Commerce	E-commerce form of business conversation and transaction based on virtual network platform	Taobao,Tmall,JD Mall, Suning,Vipshop
2	Vertical E-Commerce	Focusing on a certain industry or market segment is known as the "exclusive store" of the e-commerce industry	Red Baby Mall,Vipshop, Jumei You Pin, Letao, Mai baobao
3	Cross-Border E-Commerce	Targeted consumer groups, characterized by improving quality, optimizing operation process, improving supply chain and expanding differentiation	Tmall global, JD global, Netease koala, Vipshop, Yunji, Onion OMALL

Wechat Business(We Business) and live streaming are the new business form in the evolution of e-commerce, an online channel. Both are new forms of e-commerce. Compared with traditional e-commerce, they have the following advantages: low cost,convenient communication. In particular, the live streaming also has the advantages of wide audience, strong intuitive sense, large discount and so on(Table 16-3).

Table 16-3　E-Commerce, Wechat Business and Live Streaming

No.	Name	Advantages	Target	Supporter	Feedback	Payment	Subject
1	E-Commerce	High credibility	Sell products, attract customers	Taobao, other platforms	Store's credit rating and product evaluation	Alipay	Large commercial companies
2	Wechat Business	Low entrepreneurial costs and convenient communication	Sell products, developing agents	APP, platforms	No real and effective product evaluation	No guarantee of third-party or platform supervision	Individual or micro enterprises
3	Live Streaming	Low cost, wide audience, strong intuition, and high discounts	Sell products, building trust	Tiktok, other platform	Public praise, user evaluation of live broadcast room and order evaluation	Guarantee of third-party or platform supervision	Platforms

Part 3　How to Improve Foreign Trade Performance

As a foreign trade salesman, if you want to improve your foreign trade performance, you should not only be familiar with various foreign trade platforms, but also do the following four points:

(1) Let overseas customers find you by improving the product window, setting effective core or long tail keywords, and making effective use of data housekeeper.

(2) Let overseas customers know you by increasing the number of hits through the product title, picture click through rate, transaction conditions and attributes, the popularity of the product itself, certification and identification.

(3) Attract overseas customers to buy by improving the product details page, highlight the product advantages, improve the praise rate.

(4) Conclude orders by building trust and maintain customer relationships.

New Words and Expressions

1. abbreviate　缩写
2. enterprise　企业
3. convenience　便利；方便
4. low-cost　低成本
5. financial　金融的；财政的
6. epidemic　疫情
7. blowout　爆发
8. channel　渠道
9. mature　成熟
10. stage　阶段
11. platform　平台
12. online　线上
13. offline　线下
14. boundary　边界；界限
15. blur　模糊
16. consumption　消费
17. demand　需求
18. vertical　垂直的
19. cross-border　跨境
20. constraint　约束
21. large-scale　大型；大规模
22. third-party　第三方
23. transaction　交易
24. virtual　真实的
25. exclusive store　专卖店
26. target　目标
27. live streaming　直播
28. intuitive sense　直观感
29. discount　折扣；优惠
30. credit rating　信用评级
31. product evaluation　商品评价
32. agent　代理；代理商
33. guarantee　保障
34. public praise　口碑分
35. overseas　海外
36. window　橱窗
37. core　核心
38. long tail keywords　长尾关键词

39. housekeeper　管家

40. title　标题

41. click　点击

42. detail　细节

Exercises

List the advantages and disadvantages of offline and online sales in garment enterprises.

第十六章 服装电子商务

第一节 服装电子商务

一个传统产业，一个新兴产业，服装和电子商务在经过多年发展之后，已经建立起一些相当自然的联系。

电子商务源于英文Electronic Commerce，简写为EC。服装电子商务作为服装企业营销手段之一，由于其便利性、低成本和传播快越来越受到服装企业的重视。

中国的服装电子商务经历了三次大的发展机遇：第一次是2003年的"非典"疫情；第二次是2008年爆发的金融危机；第三次是2020年。传统服装行业在网络这一新型销售渠道中更呈井喷式发展，服装电商进入成熟期。

第二节 服装电子商务平台

服装电子商务从最初以B2B电子商务为主，从而带动了B2C和C2C电子商务的发展。当年批批吉服饰（上海）有限公司（PPG）将传统服装零售和电子商务结合，开创了男装B2C直销的新模式，填补了当时男装电子商务的空白，比较有名的有凡客、麦网等。现如今，服装电商平台呈现出以下趋势：B和C之间的边界越来越模糊，出现了B2B2C、C2C2B等；C2B这种定制化的形式随着消费的个性化，需求量日益增加（表16-1）。

表16-1 常用服装电商平台

序号	缩写	全称	代表
1	B2C	商家—顾客	京东、天猫、唯品会、淘宝商城、亚马逊、凡客诚品
2	B2B	商家—商家	1688、阿里巴巴国际站
3	C2C	顾客—顾客	淘宝网、拍拍

对实力较强的外贸企业，通过垂直电商和跨境电商形式，可以摆脱大型第三方跨境电商平台的约束，根据自身外贸业务的特点和发展需要，建设和运营属于自己的跨境电商平台，它们具有以下几个方面的优势：降低成本、业务专业化、满足个性化需求等（表16-2）。

表16-2　"平台电商""垂直电商"和"跨境电商"

序号	名称	平台特点	代表
1	平台电商	依托虚拟网络平台进行业务交谈、交易的电子商务形式	淘宝网、天猫商城、京东商城、苏宁易购、唯品会
2	垂直电商	专注于某一行业或细分市场，被称为电商界的"专卖店"	红孩子、唯品会、聚美优品、乐淘、麦包包
3	跨境电商	精准目标消费群，以提升质量、优化运营流程、改善供应链、扩展差异化为特点	天猫国际、京东全球购、网易考拉、唯品会、云集、洋葱OMALL

微商（英文是We Business，全民创业）和直播是电子商务这种线上渠道在业态演进过程中出现的一种新商业形态。两者都是电子商务的新形式。和传统的电商比起来，它们有着以下优势：创业成本低、沟通便利。尤其是直播，还具有受众广、直观感强、优惠幅度大等优点（表16-3）。

表16-3　电商、微商和直播

序号	名称	优点	产品目标	依附体	反馈机制	支付机制	主体
1	电商	信誉度高	销售产品，吸引客户	淘宝等平台	消费者能看到店铺的信用评级和商品评价	有第三方或平台监管的保障	大型商业公司
2	微商	创业成本低、沟通便利	销售产品，发展代理商	APP或平台	买家看不到真实有效的产品评价	没有第三方或平台监管的保障	个体或小微企业
3	直播	成本低、受众广、直观感强、优惠大	销售产品、建立信任	抖音等平台	直播带货口碑、直播间用户评价、订单评价	有第三方或平台监管的保障	各大平台

第三节　如何提升外贸业绩

作为一名外贸业务员，想要提升外贸业绩，除了要熟悉各种外贸平台外，还要做到以下四点：

（1）通过提升产品橱窗、设置有效的核心或者长尾关键词、有效利用数据管家，让海外客户找到你。

（2）通过产品的标题、图片点击率、交易条件和属性、产品自身的热度、认证与标识等提高点击量，让海外客户了解你。

（3）通过提升产品详情页，突出产品优势，提高好评率，吸引海外客户购买。

（4）通过建立信任和维系客户关系，成交订单。

Communication in Garment Foreign Trade
服装外贸交流

Chapter 17　Fashion News
第十七章　时装快讯

项目名称：时装快讯

项目内容：1. 香云纱

2. 元宇宙与虚拟时尚

3. GUCCI春/夏系列时装秀

4. 内衣元素

教学学时：3课时

教学目的：培养学生阅读、翻译英文时尚快讯的能力，提高学生
服饰审美能力以及对流行趋势把握的能力。

教学方式：由教师通过课前导入任务引出本章内容。结合当前
时尚快讯及流行元素，对英文时尚快讯进行阅读和
翻译。

教学要求：（1）掌握本课词汇。

（2）具备阅读和翻译英文时尚快讯的能力。

（3）具备服装审美以及时尚潮流的洞察力。

Chapter 17   Fashion News

Part 1   Gambiered Canton Silk

In fashion, vintage elements have always been the favorite of world' s trend. The beauty and temptation that the vintage cheongsam made of Gambiered Canton Silk bring to women is irresistible. After hundreds of years , although the Gambiered Canton Silk is ancient, it is like a new soul. Gambiered Canton Silk, retro color matching thick and fresh,smooth and delicate but has a concave convex feel. Full colors match brightly with texture of yarn , which makes the clothes fashionable and vintage with the simple, neat one buckle and attractive short slits. Admittedly, inheritance and innovation are the two major concepts of fashion design. The vintage Gambiered Canton Silk perfectly combines the characteristics of cheongsam and speciality of silk, perfectly realizing the integration of classic and fashion. It not only conveys beauty and confidence by breathe, but also inherits intangible cultural heritage.

Part 2   Metaverse and Virtual Fashion

In recent years, the concept of the metaverse has gradually entered the public view. In order to cater to the new generation of young consumers, the fashion industry speeds up its digital transformation through the new medium of the Metaverse. In addition to digital marketing and new online sales channels, a new economic form—virtual fashion, has been derived. Through 3D modeling, printing and other latest techniques and expressions, 3D digital fabrics and garment have been developed. People can interact with garment, feel the changes in color and shape of garment via digital technology, which makes people own a wonderful digital sensory experience while appreciating the garment. What' s more, by means of the Metaverse, garment enterprise have a better insight into consumers' cognition and experience of garment, creating future with consumers in terms of styles, trends, colors, etc, and explore a lifestyle that covers new cultural core and virtual non- homogeneous commercial value.

Part 3　GUCCI Spring/Summer Collection

GUCCI Spring/Summer 2019 fashion show was held in Paris, France. The strong retro elements are still characterized with the iconic design of Spring/Summer Collection. The elegant long skirts, pleats, batwing sleeves and curved sleeve designs fully demonstrate the retro fashion. Suits are paired with classic patterns such as Prince of Wales plaid, which gives the suits a sense of leisure and carries their own personality as well.

The color matching and application is definitely a highlight. The application of pure white shows the designer's exquisite design, manifesting noble quality. Winter jasmine's yellow, purple, lapis blue, green, red, lake blue and other high-saturation colors have super visual impact, and even if they are contrasting colors, they are more attractive due to different materials and textures. The texture and design of fruit printing make GUCCI gradually become younger, and constantly express the emotion of pursuing independence and yearning for freedom.

Part 4　Underwear Elements

Gorgeous and fashionable underwear is increasingly grabbing people's attention in fashion T stage. The most eye-catching fashion elements of this season show its unique attractive magic. The gorgeous angel decoration is only used to set off the unique denim elements. The classic denim elements are disrupted and spliced to become a new fashion trend. The bright and unique plaid pattern is used in the new underwear design, which adds both a sense of youth and a POP trendy impression.

Sexy graffiti elements have evolved into oversized letter patterns in underpants, looking cute and charming and they are simple and yet have a sense of design to bring out the youthful vigor of beautiful girls. The casual smoky gray stitching fully expresses sense of style, which is convenient for cool girls' summer matching.

New Words and Expressions

1. vintage　老式的；古色古香的
2. gambiered canton silk　香云纱
3. irresistible　无法抗拒的；无法抵制的
4. cheongsam　旗袍
5. concave-convex　凹凸的
6. short slit　开衩短裙
7. inheritance　继承物；遗产；遗传特征
8. collection　时装展览；时装发布会
9. retro　怀旧的；重新流行的
10. pleat　活褶；裤褶

11. batwing sleeve　蝙蝠袖
12. leisure　闲暇；休闲活动
13. highlight　突出；强调；使显著；加亮
14. exquisite　精致的；精美的
15. manifesting　使显现；显示
16. winter jasmine　迎春花；冬茉莉
17. lapis blue　青金石蓝
18. texture　质地；纹理
19. pursuing　从事；追赶；追求
20. yearning　怀念；渴望
21. metaverse　元界；元宇宙(虚拟空间)
22. virtual　虚拟的；模拟的
23. concept　概念；观念
24. cater to　迎合；为……服务
25. digital　数字的；数码的
26. derive　从……衍生出；源于

27. fabric　纤维织物；面料
28. sensory　感觉的；感官的
29. cognition　认识；认知；认识的结果
30. core　果核；中心部分；核心
31. non-homogeneous　非齐次的；非均质的
32. commercial　商业的；商务的
33. trend　趋势；动态；时尚
34. philosophy　哲学；理念
35. costume　服装；装束
36. convey　传送；运输；表达；传递
37. grab　引人注意；吸引
38. eye-catching　引人注目的；耀眼的
39. denim　斜纹粗棉布；丁尼布
40. plaid　格子花呢；格子图案
41. graffito　涂鸦
42. oversize　太大的；超大型的；特大号

Exercises

Translate the following passage into Chinese.

Sexy graffiti elements have evolved into oversized letter patterns in underpants, looking cute and charming and it is simple and yet has a sense of design to bring out the youthful vigor of beautiful girls.

导入：你知道当下流行趋势吗？

第十七章　时装快讯

第一节　香云纱

在服装时尚中，复古元素一直是世界时尚的宠儿。复古香云纱旗袍带给女性的美丽和诱惑是让人难以抵挡的。经历了几百年的沉淀，香云纱虽古老，却如同新生的灵魂。香云纱用于制作旗袍的面料，复古的配色厚重而不失清新，光滑而细腻却有着凹凸有致的手感，饱满的花色与香云纱绸缎肌理相得益彰，简约整齐的一字扣，漂亮的短开衩，时尚而又复古的艺术之美极尽呈现。传承与创新，是服装设计的两大理念。复古香云纱旗袍，恰到好处地融合了旗袍和香云纱的特性，完美地实现了经典与时尚一体，把美与自信传递在一呼一吸之间，把非遗文化镌刻在细细密密的纹路里。

第二节　元宇宙与虚拟时尚

近年来，元宇宙的概念逐渐进入大众的视野。为迎合新一代年轻人的消费市场，时尚产业也通过元宇宙这一新媒介，加快数字化转型。除了数字营销、线上销售新渠道，还衍生出了一类全新的经济形态——虚拟时尚。通过3D建模、打印等最新的技艺和表现手法，研发出3D数字面料和服饰，通过数字技术，人能够与服装进行互动，感受服装颜色和形态的改变，在欣赏服装的同时，也是一场美妙的数字化感官体验。通过元宇宙，服装企业可以更好地洞察消费者对服装的认知和体验，从款式、潮流、色彩等方面与消费者共创未来，与消费者共同探索一种涵盖新的文化内核、虚拟的非同质化商业价值的生活方式。

第三节　GUCCI春/夏系列时装秀

GUCCI 2019春夏系列时装秀在法国巴黎举行。浓郁的复古元素依旧是春夏系列中标志性的设计特色，优雅的长裙、褶皱、蝙蝠袖以及弧形袖设计，充分展现出复古时尚。西装套装搭配经典纹样，例如威尔士亲王格纹，既赋予西装一丝休闲又承载着自我的个性。

色彩的搭配与应用绝对是一大亮点，纯白色调的应用展现出设计师的精湛设计，彰显出高贵的品质。迎春花黄、紫、青金石蓝、绿、红、湖蓝等多种高饱和度的色彩拥有超强的视觉冲击，即使是撞色，也因不同的材质和纹理更具魅力。水果印花的纹理与设计让GUCCI逐渐走向年轻化，不断表达追求独立、向往自由的情感。

第四节　内衣元素

在时尚T台上，华丽时尚的内衣越来越受到人们关注。本季最养眼的流行元素展示出其独有的吸睛魔力。华丽的天使装饰只为衬托独特的牛仔元素。将经典牛仔元素打乱拼接，成为新的时尚潮流。明亮独特的格纹图案用于新的内衣设计中，不仅增加年轻感，还给人一种POP的新潮印象。

性感俏皮的涂鸦元素演变成超大号的字母图案，用于内裤的设计尤其可爱吸睛，简单而不失设计感，衬托出美丽女孩的青春活力。率性的烟灰色拼接充分表现时尚的硬朗一面，便于酷感女孩的夏日搭配。

本模块微课资源（扫描二维码观看视频）

14.1　Bargain in Garment Foreign Trade

15.1　Principles of Professional English Translation in Garment

16.1　认识微商、电商和直播

16.2　如何提升外贸业务？

17.1　Cheongsam

Appendix I   Fiber and Fabric
附录1   纤维与织物

Fibers   纤维

antistatic fiber   抗静电纤维

animal Fiber   动物纤维

acetate （cellulose）fiber   醋酯纤维

chemical fiber   化学纤维

gold fiber   金属纤维

jute   黄麻

mohair   马海毛

nylon   尼龙

synthetic fiber   合成纤维

staple   短纤维

triacetate   三醋酸纤维

viscose fiber   黏胶纤维

vinylon   维尼纶

ceramic fiber   陶瓷纤维

regenerated fiber   再生纤维

camel hair   骆驼毛

filament   长纤维

hemp   大麻

kapok   木棉

artificial fiber （man-made Fiber）   人造纤维

polyester fiber   涤纶（聚酯纤维）

viscose rayon   黏胶人造丝

viscose acetate fiber   黏胶醋酯纤维

acrylic fiber   腈纶

polyamide fiber   锦纶（聚酰胺纤维）

Fabric   织物

asbestos cloth   石棉布；石棉织物

aberdeen fabric   阿巴丁布

acrylic   腈纶

cotton   棉布

silk fabric   丝织物；绸缎

canvas   马尾衬布；帆布

cashmere   开司米织物，精纺毛针织物

crepe   绉布

cerecloth   蜡布；胶布；漆布

chiffon   雪纺

denim   牛仔布（又称劳动布）

duck   帆布

fur   皮革

flameproof fabric   防火布

gray clothes　坯布

herringbone twill　人造斜纹布

knit fabric　针织布

lycra　莱卡

leather　革皮

linen　麻布

linen like fabric　仿麻织物

loop pile　毛巾布

melton　麦尔登呢（领底绒布）

non-woven fabric　非织造布

figured cloth　提花布

yarn dyed fabric　色织布

drill　斜纹布

khaki drill　卡其布

serge　哔叽

gabardine　华达呢

mesh fabric　网眼布

burnt-out fabric　烂花布

cotton/polyester fabric　涤/棉织物

bast fabric　麻织物

flannelette　法兰绒绒布

seersucker　泡泡纱

poplin　府绸

tussores　罗缎

haircords　麻纱

velveteen　平绒

corduroy　灯芯绒

plain cloth　平布

cotton fabric　棉织物

habotai　电力纺

nylonpalace　尼龙纺

huachun habotai　华春纺

fuchun habotai　富春纺

crepe de chine　双绉

kabe crepe　璧绉；金丝双绉

crepe georgette　乔其绉（又称乔其纱）

polyester crepe　涤丝绉

velvet　绒

wool fabric　毛织物

flannel　法兰绒

nylon　尼龙布

rayon　人造丝

silk　真丝

satin/sateen　缎纹（色丁）

taffeta　塔府绸

tweed　粗花呢，毛绒布

towel cloth　毛巾布

waterproof fabric　防水布

oxford　牛津布

twine cloth　线呢

silk-like fabric　仿丝织物

spun-like fabric　仿短纤纱型织物

wool-like fabric　仿毛织物

Appendix II   Garment Accessory

附录2   服装辅料

belt   腰带；皮带；带
button   纽扣
animal bone button   动物骨纽扣
bow   蝴蝶结
binding tape   滚条（绲条）；绑带
buckle loop   纽圈；金属扣襻
collar stay   领插襻竹
cotton string   棉绳
care label   洗水唛
elastic   橡皮筋；松紧带
eyelet   网眼
eyes & hooks   钩扣；风纪扣
interlining   衬料
fusible interlining   黏合衬
non-fusible interlining   非黏合衬
lace   花边；蕾丝
lining   里布
main label/brand name   主唛；商标
label cloth   商标带

rivet   包头钉；铆钉
ribbing   罗纹；罗口
shoulder pad   垫肩
stripe   带条；绳子
snap   按扣；揿纽
size label   尺码唛
thread   线
velcro   魔术贴；尼龙搭扣
chain zipper   链齿式拉链
yarn   纱线
sewing thread   缝纫线
embroidery thread   绣花线
hand knitting yarn   手编绒线
bast thread   麻线
lame yarn   金银线
cordage   绳
narrow fabric   带织物
ceramic button   陶瓷纽扣

Appendix III　Production Terms

附录3　生产术语

Cutting Room　裁剪车间

bundling　捆扎
bias cut　斜纹；斜丝缕；裁剪
basic block　基本纸样
cutting　开剪；裁布
cutting pieces　碎料；裁片
gross cut　横纹（横丝）裁剪
C.M.T.（cutting，making，trimming）裁剪、缝制、整装（简称裁、缝、整）
drilling　钻孔位
drafting　绘图
grading　放码；推板
grain　布纹；丝缕
handling　处理；手感
layout　排料
marker　唛架；排料图
assortment　分色分码
production pattern　生产纸样
preshrinking　预缩水
plaid matching　对格
spreading　拉布；铺布
size specification　尺码表
splicing　驳布位；拼接位
shading　色差
fabric sample（swatch）　布板
straight cut　直纹（直丝）裁剪
sorting　分码
tolerance　放松度；抛位
tubular fabric　圆筒形布圈
pattern　纸样

Sewing Room　缝制车间

attach collar　绱领
blinding　挑脚
blind stitch　暗缝
buttoning　钉纽扣
button holing　开扣眼
basting　假缝；粗缝；绗缝
binding/bound　绲边；包缝
bundle system　捆扎系统
bartack　套结；打枣
back stitch　回针
break stitch　断线
break needle　断针

chain stitch   链式线迹

cut & sewn   裁缝（切驳）

clean finish   光洁整理；还口

cross crotch   十字缝；裆底缝

crease line   裤缝线；折线；裤中骨

crotch   裤裆；裤浪；裤大小裆弯

covering stitch   覆盖线迹

dart   省道；死褶

double needle fell seam   双线埋夹

edge stitching   缝边线迹

elastic waistband   松紧腰带

easing   吃势；容位

embroidery   绣花

fullness   宽松量

flow chart   流程表

flat seam   平缝

fitting   试衣；试身

fusing interlining   烫黏合衬

final inspection   终检

fold back facing   连衣门襟（原身出贴）

forepart   前片部分

hanger   衣架

handling time   处理时间

hemming   卷边

in-process inspection   中检

ironing   熨烫

iron-shine   熨反光（起镜面）

inspection   检查（查衫）

join crotch   缝合裆位（埋浪）

labour cost   劳工成本

looping   做裤襻（起耳仔）

M/C maintenance   机械保养

material   物料

material handling   物料输送

missed stitch   漏针

notch   剪口；扼位；刀眼

overhead   经常费；营业费用（厂皮）

open seam   劈缝（开骨）

overtime working   超时工作

operation break down   分科

overlap   重叠（搭位）

one layer yoke   一片育克（过肩）

overlock/serging   锁边/包缝/及骨

overlock with 5 threads   五线锁边

off pressing   终烫

piping   绲边（捆条）

pleat   褶

production schedule   生产进度表（生产排期）

piece rate   计件工资

puckering   起皱

pressing   熨烫

Q.C.（quality control）   品质控制

quilting   绗缝（间棉）

quota   配额

run stitch   初缝线迹（运线；车线）

stitch   线迹（针步）

seam   缝骨；缝型；缝口

set in sleeve   绱袖

supervisor   主任（主管）

stay tape   胸衬条

skipped stitch   跳线

sewing waistband with waistband machine   拉腰头

S.A.（seam allowance）   缝头（子口）；缝份

top stitching   面缝线迹（压明线）

tucking   排褶车缝

turned finish   翻折处理（还口）

two layer yoke   两片育克（过肩）

trimming   剪线

W.I.P.（work in progress）   半成品

under pressing   中烫

wrinkles   不平服；不平整

waistband is extension of body   连衣腰头（原身出腰头）

zig-zag stitch   人字线迹

packing   包装

Appendix IV   Clothing Industry Tools

附录 4   制衣设备

model   模特；人台
tracer   描线器；轨迹轮
notcher   记号剪；刀眼机；打扼机
puncher   打孔机
multi-grader   尺码放缩仪
hand shear   剪刀
drill   锥子
meter rule   米尺
un-picker   拆线器
pin   大头针
sniper   剪线刀
tape measure   软尺
chalk   划粉
straight knife   直剪；直刀式裁剪机
rotary knife   圆刀式裁剪机
bend knife   带刀式裁剪机
marker copier   唛架复制机
single needle lockstitch machine   单针平缝机
double needles lockstitch machine
双针平缝机

multi-needle lockstitch machine
多针平缝机
three threads overlock machine
三线包缝机；锁边机
five threads overlock machine   五线包缝机
flat lock machine   平缝机
bartacker   套结机
button holing machine   开扣眼机
buttoning machine   钉扣机
chain stitch machine   锁链机
blind stitch machine   暗缝机
electrical iron   电熨斗
steam-electrical iron   蒸汽熨斗
collar-cuff-flap former   烫领袖口机
collar turner   翻领机
tuck machine   压褶机
pleating machine   压褶机

Appendix V   Colors
附录5   颜色

red   红色

pink   粉红

baby pink   浅粉红

carmine   紫红

garnet   暗红；酱红

wine red   葡萄酒红

geranium   原色红

rose   玫瑰红

French rose   法国红

red exide   褐红

copper red   铜色

amber   琥珀色

maroon   茶色；栗色

sandy   淡茶色

vmber   浓茶色

pale   土色

coco   黄棕色

khaki   卡色

olive   橄榄色

purple   紫色

violet   紫罗兰

lilac   淡紫色

lavender   淡紫色

yellow   黄色

buff   浅黄色；米色

canary   淡黄色

deep yellow   深黄

golden yellow   金黄

york yellow   蛋黄

onix   朱黄

bluish yellow   青黄

citrine   柠檬黄

orange   橘黄

blue   蓝色

baby blue   浅蓝

zircon   浅蓝

Prussian blue   普鲁士蓝

sky blue   天蓝

navy blue   海军蓝

hyacinth   紫蓝；紫丁香色

cyanine blue   原色蓝

sea blue   海蓝

calamine blue   浅蓝

manadarin blue   深蓝

royal blue   宝蓝

lyons blue   蓝紫；里昂蓝

ming blue   藏青

blue gray   蓝灰

cream   米色

coral   珊瑚色

purple bronze   紫铜色

copper color   古铜色

brown　棕色

beige　灰棕色

tan　浅棕色

gray　灰色

dark gray　深灰

bay　枣色

white　白色

off-white　灰白；朱白

green　绿色

viridesent　淡绿色

pea green　豆绿

olive green　橄榄绿

celadon　灰绿

dark green　深绿

deep green　墨绿

black　黑色

gold　金色

silver　银色

pied　杂色

transparent　透明

translucent　半透明

opaque　不透明